LiDAR Remote Sensing
and Applications

Remote Sensing Applications

Series Editor
Qihao Weng
Indiana State University
Terre Haute, Indiana, U.S.A.

LiDAR Remote Sensing and Applications

Pinliang Dong and Qi Chen

CRC Press
Taylor & Francis Group
Boca Raton London New York

CRC Press is an imprint of the
Taylor & Francis Group, an **informa** business

CRC Press
Taylor & Francis Group
6000 Broken Sound Parkway NW, Suite 300
Boca Raton, FL 33487-2742

© 2018 by Taylor & Francis Group, LLC
CRC Press is an imprint of Taylor & Francis Group, an Informa business

No claim to original U.S. Government works

Printed on acid-free paper

International Standard Book Number-13: 978-1-4822-4301-7 (Hardback)
International Standard Book Number-13: 978-1-138-74724-1 (Paperback)

Library of Congress Cataloging-in-Publication Data

Names: Dong, Pinliang, author. | Chen, Qi (Geography professor), author.
Title: LiDAR remote sensing and applications / Pinliang Dong and Qi Chen.
Description: Boca Raton, FL : Taylor & Francis, 2018. | Includes
bibliographical references.
Identifiers: LCCN 2017035053 | ISBN 9781482243017 (hardback : alk. paper)
Subjects: LCSH: Remote sensing. | Geography--Remote sensing. | Forests and
forestry--Remote sensing. | Ecology--Remote sensing. | Optical radar.
Classification: LCC G70.4 .D66 2018 | DDC 621.36/78--dc23
LC record available at https://lccn.loc.gov/2017035053

Visit the Taylor & Francis Web site at
http://www.taylorandfrancis.com

and the CRC Press Web site at
http://www.crcpress.com

Contents

LiDAR Data Index Map

Base map from Esri ArcGIS online.

Note: A study area in Ghana for biomass estimation (p. 89) is not shown on this map.

Foreword

I am very pleased to write the foreword for *LiDAR Remote Sensing and Applications* authored by Prof. Pinliang Dong and Prof. Qi Chen, who are among leading experts in the field. With over 35 years of research experience in remote sensing and digital earth, I have witnessed extraordinary accomplishments of Earth observation from spaceborne, airborne, and ground-based platforms, using multispectral, hyperspectral, radar, and light detection and ranging (LiDAR) instruments. Remote sensing has greatly improved our understanding of the natural and built environments and human-environment interactions. With its unique capabilities for collecting highly accurate three-dimensional coordinates of objects, LiDAR has been widely used in many areas including vegetation mapping, urban studies, and geosciences. I am glad that the authors have taken keen interest in writing a reader-friendly book on the subject.

This is a unique book in that it smoothly combines LiDAR principles, data processing methods, applications, and hands-on practices, following an overview of remote sensing. An index map of LiDAR data and a list of abbreviations are also included to improve the readability of the book. For forest applications, readers can find examples such as creating leaf-on and leaf-off canopy height models in Susquehanna Shale Hills, PA; identifying disturbances from lightning and hurricane in mangrove forests in Florida; and estimating aboveground biomass in tropical forests in Ghana. For urban applications, readers can see examples such as road extraction, powerline corridor mapping, and population estimation in Denton, TX; parcel-based building change detection in Surrey, Canada; and road blockage detection in Port-au-Prince after the 2010 Haiti earthquake. For geoscience applications, readers can explore samples such as measuring dune migration rates in White Sands, NM; analysis of offset channels associated with the San Andreas Fault in California; and trend surface analysis and visualization of rock layers in Raplee Ridge, UT. Undergraduate and graduate students will find that the 11 step-by-step GIS projects with LiDAR data can really help them understand LiDAR data processing, analysis, and applications, while professionals and researchers will benefit from various topics on LiDAR remote sensing and applications, along with over 500 references in the book.

I'd like to congratulate Prof. Dong and Prof. Chen on their achievements. Although no book can convey every detail in a field or discipline, *LiDAR Remote Sensing and Applications* contains enough information for undergraduate/graduate students, professionals, and researchers, and is worth reading more than once, in my humble opinion.

Guo Huadong
Academician, Chinese Academy of Sciences
President, International Society for Digital Earth (ISDE)
Editor-in-Chief, International Journal of Digital Earth
Beijing, China
June 2017

Preface

The last decade has seen a rapid increase in the applications of light detection and ranging (LiDAR) in various fields. This book introduces the fundamentals of LiDAR remote sensing, LiDAR data processing, and LiDAR applications in forestry, urban environments, and geosciences. LiDAR data collected in 27 areas in the United States, Brazil, Canada, Ghana, and Haiti are included in the book, and a total of 183 figures were created to introduce the concepts, methods, and applications in an easily understood manner, along with over 500 references. Compared with some other books on LiDAR, a unique feature of this book is the combination of LiDAR principles, data processing basics, applications, and hands-on practices. The 11 step-by-step projects are mostly based on Esri's ArcGIS software to support seamless integration of LiDAR products and other GIS data. Over 4.4 GB of LiDAR data for the projects are available online, fully tested in ArcGIS 10.2.2 and 10.4.1, and can be used for ArcGIS 10.2 and later versions. The first six projects are for basic LiDAR data visualization and processing, while the remaining five projects cover more advanced topics: mapping gaps in mangrove forests in Everglades National Park, FL, analyzing powerline corridor in Denton, TX, estimating small-area population in Denton, TX, measuring sand dune migration in the White Sands Dune Field, NM, and generating trend surfaces for rock layers in Raplee Ridge, UT.

This book includes many references to recent studies, and can be used as a textbook or reference book by undergraduate and graduate students in the fields of geography, forestry, ecology, geographic information science, remote sensing, and photogrammetric engineering. The hands-on projects are designed for under-graduate and graduate students who have worked with vector and raster data in ArcGIS. The questions after Projects 4.1, 4.2, 5.1, 5.2, 6.1, and 6.2 can be used by instructors as homework or project assignments for senior undergraduate or graduate students. Professionals in industry and academia will also find this book useful.

We would like to thank the following entities and researchers for providing LiDAR and other data in the book: The OpenTopography facility based at the San Diego Supercomputer Center and supported by the National Science Foundation (NSF), the National Center for Airborne Laser Mapping (NCALM) funded by NSF for data collection through various projects, the Ministry of Economy, Trade, and Industry (METI) of Japan, the United States National Aeronautics and Space Administration (NASA), Rochester Institute of Technology, Kucera International (under sub-contract to ImageCat Inc. and funded by the Global Facility for Disaster Reduction and Recovery (GFDRR) hosted at The World Bank), the U.S. Census Bureau, the U.S. Geological Survey, the São Paulo Research Foundation (FAPESP, Brazil), City of Surrey (British Columbia, Canada), the IndianaMap Framework Data, the Texas Natural Resources Information System (TNRIS), Esri, Cheng Wang, Mehmet Erbas, Gherardo Chirici, Davide Travaglini, and Krzysztof Stereńczak. We also thank our students, colleagues, and friends for their encouragement and support. Last but not least, we wish to thank our wives, children, and parents for their love and support, and Pinliang's wife Yijin for designing the book cover.

Pinliang Dong

Qi Chen

Authors

Dr. Pinliang Dong is a Professor in the Department of Geography and the Environment, University of North Texas (UNT), Denton, TX, USA. He received his B.Sc. in geology from Peking University, China in 1987, M.Sc. in cartography and remote sensing from the Institute of Remote Sensing Applications, Chinese Academy of Sciences in 1990, and Ph.D. in geology from the University of New Brunswick, Canada in 2003. Before joining UNT in 2004, he worked as a Senior GIS Analyst/Programmer at Titan Corporation in California (USA) and Okinawa (Japan), and a Staff Associate/GIS Specialist at the Center for International Earth Science Information Network (CIESIN), Columbia University. His research interests include remote sensing, geographic information systems (GIS), digital image analysis, and LiDAR applications in forestry, urban studies, and geosciences. He has taught Intermediate GIS, Advanced GIS, Advanced GIS Programming, Remote Sensing, Special Topics in GIS: LiDAR Applications, and China Field School; mentored over 30 Master's and doctoral students and two post-doctoral fellows; and hosted over 15 international visiting scholars. He is a member of the American Association of Geographers (AAG), American Geophysical Union (AGU), and International Society for Digital Earth (ISDE).

Dr. Qi Chen is a Professor in the Department of Geography at the University of Hawaii at Mānoa, Honolulu, Hawaii, USA. He received his B.Sc. and M.Sc. in Geography in 1998 and 2001, respectively, from Nanjing University, China, and Ph.D. in Environmental Sciences, Policy, and Management from the University of California, Berkeley, USA in 2007. He joined the University of Hawaii at Mānoa as a tenure-track Assistant Professor in 2007, was tenured and promoted to Associate Professor in 2012, and was promoted to Professor in 2017. His early interest and research in LiDAR remote sensing during his Ph.D. study were to develop effective methods for LiDAR data processing and information extraction, including airborne LiDAR point cloud filtering, digital terrain model generation, and individual tree mapping. His research in recent years has expanded to satellite LiDAR, terrestrial LiDAR, and the use of LiDAR for extracting various vegetation attributes for applications such as wildfire mapping and hazard analysis, biodiversity and habitat analysis, and biomass mapping and estimation. His overall interest in LiDAR remote sensing is to improve the methods of information extraction from LiDAR and to promote the use of LiDAR for assisting environmental management and decision making. He is an advisor to many master's and doctoral students, postdocs, and has hosted many international students and professors for studying LiDAR in his research lab. He has also given multiple tutorial workshops on LiDAR remote sensing in international conferences organized by professional societies.

List of Abbreviations

AGB	Aboveground Biomass
AIS	Airborne Imaging Spectrometer
ALS	Airborne Laser Scanning
ALSM	Airborne Laser Swath Mapping
AOR	Angle of Repose
ASCII	American Standard Code for Information Interchange
ASPRS	American Society for Photogrammetry and Remote Sensing
ASTER	Advanced Spaceborne Thermal Emission and Reflection Radiometer
ATM	Airborne Thematic Mapper
AVIRIS	Airborne Visible and Infrared Imaging Spectrometer
CBERS	China-Brazil Earth Resource Satellite
CHM	Canopy Height Model
CMM	Canopy Maximum Model
CRM	Component Ratio Method
DBH	Diameter at Breast Height
dDSM	Differenced Digital Surface Model
DEM	Digital Elevation Model
DGPS	Differential Global Positioning System
DHM	Digital Height Model
DOQQ	Digital Orthophoto Quarter Quadrangle
DSM	Digital Surface Model
DTM	Digital Terrain Model
ENP	Everglades National Park
EO-1	Earth Observing-1
ETM+	Enhanced Thematic Mapper Plus
EVLR	Extended Variable Length Records
FIA	Forest Inventory Analysis
FiND	Fishing Net Dragging
FGDC	Federal Geographic Committee
GDEM	Global Digital Elevation Map
GIS	Geographic Information System
GLAS	Geoscience Laser Altimeter System
G-LiHT	Goddard's LiDAR, Hyperspectral and Thermal Imager
GPS	Global Positioning System
GWR	Geographically Weighted Regression
HLC	Height to Live Crown
HRV	High Resolution Visible
HUM	Housing Unit Method
ICESat	Ice Cloud and Land Elevation Satellite
IDW	Inverse Distance Weighting
IMU	Inertial Measurement Unit
InSAR	Interferometric Synthetic Aperture Radar

INU	Inertial Navigation Unit
ISODATA	Iterative Self-Organization Data Analysis Technique
LAI	Leaf Area Index
LiDAR	Light Detection and Ranging
LVIS	Laser Vegetation Imaging Sensor
MBLA	Multi-Beam Laser Altimeter
MHC	Multiresolution Hierarchical Classification
MMO	Mathematical Morphological Operation
MSS	Multispectral Scanner
MTLS	Mobile Terrestrial Laser Scanning
NAD27	North American Datum of 1927
NAD83	North American Datum of 1983
NASA	National Aeronautics and Space Administration
NAVD88	North American Vertical Datum of 1988
NCALM	National Center for Airborne Laser Mapping
NEON	National Ecological Observatory Network
nDSM	Normalized Digital Surface Model
NOAA	National Oceanic and Atmospheric Administration
NSF	National Science Foundation
OBIA	Object-Based Image Analysis
OLI	Operational Land Imager
OLS	Ordinary Least Squares
POS	Positioning and Orientation System
PSTP	Pairs of Source and Target Points
RANSAC	Random Sample Consensus
RBV	Return-Beam Vidicon
RMSE	Root Mean Square Error
SAF	San Andreas Fault
SIR-A	Shuttle Imaging Radar-A
SIR-B	Shuttle Imaging Radar-B
SIR-C	Shuttle Imaging Radar-C
SLA	Shuttle Laser Altimeter
SLICER	Scanning Lidar Imager of Canopies by Echo Recovery
SPOT	Système Probatoire d'Observation de la Terre
SRS	Simple Random Sampling
STLS	Stationary Terrestrial Laser Scanning
STRM	Shuttle Radar Topography Mission
THD	Treetop Height Difference
TIMS	Thermal Infrared Multispectral Scanner
TIN	Triangulated Irregular Network
TIRS	Thermal Infrared sensor
TLS	Terrestrial Laser Scanning
TM	Thematic Mapper
TNRIS	Texas Natural Resources Information System
TPS	Thin-Plate Spline
USGS	U.S. Geological Survey

VGI	Volunteered Geographic Information
VHR	Very-High-Resolution
VLR	Variable Length Records
VNIR	Visible and Near-Infrared
WGS84	World Geodetic System of 1984
WFT	Windowed Fourier Transform
WSDF	White Sands Dune Field

1 Brief Overview of Remote Sensing

1.1 FROM AERIAL PHOTOGRAPHY TO REMOTE SENSING

Photography from aerial platforms was initially conducted using ornithopters, balloons, kites, pigeons, and gliders (Jensen 2006). In 1908, 5 years after the Wright brothers built the world's first operational aircraft, an aircraft was first used as a platform for aerial photography. During World War I and World War II, aerial photography played an important role as a military photo-reconnaissance method. In the 1920s and 1930s, aerial photography became the standard information source for the compilation of topographic maps. From the 1930s until the early 1960s, black-and-white, color, and color-infrared aerial photographs were widely utilized by geologists, foresters, and planners for interpreting the Earth's surface features (van Nowhuys 1937, Melton 1945, Desjardins 1950, Miller 1961). The use of aerial photographs improves the efficiency of many mapping applications because (1) aerial photographs make it possible for mapping ground features in areas where field investigation is difficult due to poor accessibility; (2) stereo aerial photographs help the interpretation of ground features through incorporation of topographic information; and (3) color-infrared aerial photographs provide spectral information beyond human vision. The main drawbacks of early aerial photographs were that (1) aerial photograph acquisition depended on the weather and (2) aerial photographs were normally recorded in an analog format and were not calibrated, which precludes quantitative analysis.

The term "remote sensing" was first coined by Evelyn Pruitt of the U.S. Office of Naval Research in the 1950s, and the traditional aerial photography gradually evolved into remote sensing around 1960. Sabins (1987) defined remote sensing as methods that employ electromagnetic energy to detect, record, and measure the characteristics of a target, such as the Earth's surface. Although many other definitions of remote sensing exist in literature (Colwell 1984, Fussell et al. 1986, Jensen 2006), it is commonly accepted that the basis for remote sensing is the electromagnetic spectrum (Figure 1.1). Since the late 1960s and early 1970s, many traditional aerial photographic systems have been replaced by airborne and spaceborne electro-optical and electronic sensor systems. While traditional aerial photography mainly works in visible bands, modern spaceborne, airborne, and ground-based remote sensing systems produce digital data that covers visible, reflected infrared, thermal infrared, and microwave spectral regions with different spatial, spectral, temporal, and radiometric resolutions. Traditional visual interpretation methods in aerial photography are still useful, but remote sensing encompasses more activities such as theoretical modeling of target properties, spectral measurement of objects, and digital image analysis for information extraction.

1

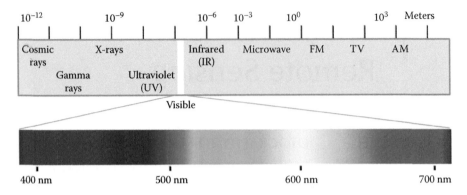

FIGURE 1.1   The electromagnetic spectrum. The numbers show wavelengths of spectral regions.

There are two types of remote sensing systems: passive and active (Figure 1.2). Passive remote sensing systems measure reflected solar radiation in visible, near-infrared, and mid-infrared wavelengths, or absorbed and then reemitted solar radiation in thermal infrared wavelengths. Active remote sensing systems, on the other hand, emit radiation toward the target using their own energy source and detect the radiation reflected from that target. An important advantage for active sensors is

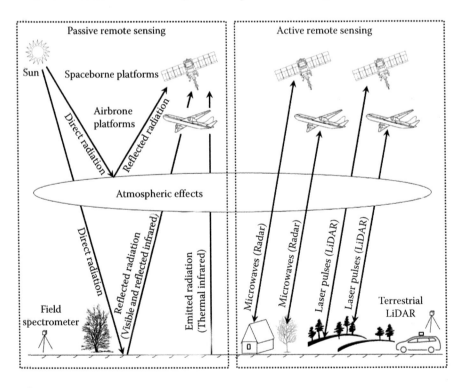

FIGURE 1.2   Passive and active remote sensing.

their ability to obtain measurements independently of sun illumination conditions and largely independent of weather conditions. The following sections provide an overview of two passive remote sensing methods—multispectral remote sensing and hyperspectral remote sensing—and two active remote sensing methods—radar remote sensing and light detection and ranging (LiDAR) remote sensing.

1.2 MULTISPECTRAL REMOTE SENSING

In multispectral remote sensing, visible and reflected infrared (near infrared and mid-infrared) images are collected by recording the reflection of solar radiation from the earth using airborne and spaceborne sensors, whereas thermal infrared images are collected by recording emitted thermal radiation from the earth. An early example of airborne visible and reflected infrared sensors is the Airborne Thematic Mapper (ATM), an eleven-band prototype of the Thematic Mapper (TM) of the Landsat-4 satellite. In additional to airborne visible and reflected infrared multispectral sensors, airborne thermal infrared sensors also provide important data for many applications, especially geologic mapping and mineral exploration. Hunt (1980) reported that silicates exhibit fundamental vibrational stretching modes in the 10 μm region. The reflection peak at or near the fundamental vibration frequency is called the reststrahlen or residual ray peak (Goetz 1989). Kahle (1984) reported that the absorption features of silicate rocks shift toward longer wavelengths with the decrease of silica content from quartzite through basalt. Since silicates make up the bulk of the crustal rocks, and the fundamental vibrational features of silicates are located in the 8–14 μm atmospheric transmission window, an emissivity minimum resulted from the reststrahlen can be detected with multispectral sensors. Kahle and Rowan (1980) used multispectral thermal data from a Bendix 24-channel scanner for lithological mapping in the East Tintic Mountains in central Utah, USA. Their study showed that it is possible to discriminate among several rock types primarily based on their silica content. Since the spectral properties of minerals may be quite different in visible and reflected infrared region, it is possible to discriminate among carbonate rocks, quartzite, quartz latitic and quartz monzonitic rocks, latitic and monzonitic rocks, silicified altered rocks, and argillized altered, if multispectral thermal data are combined with visible and reflected infrared data (Goetz 1989). The study by Kahle and Rowan (1980) provided the justification for the development of a multispectral scanner working in the thermal infrared region (Goetz 1989). In the early 1980s, the Thermal Infrared Multispectral Scanner (TIMS) was developed in the United States for remote sensing of nonrenewable resources. The TIMS instrument collects thermal emission energy in six bands near the peak of the Earth's surface emission (8.2–8.6, 8.6–9.0, 9.0–9.4, 9.4–10.2, 10.2–11.2, and 11.2–12.2 μm). Using TIMS data, Kahle and Goetz (1983) showed that it was possible to map quartz-bearing rocks. Gillespie et al. (1984) used TIMS data to map alluvial fans in Death Valley, California, and found that both composition and relative age were recognizable.

The use of airborne visible, reflected infrared, and thermal infrared sensors has a number of benefits. The user can select the wavebands of interest in a particular application, and the aircraft can be flown to specific user requirements concerning time of day, flying direction, and spatial resolution. However, data acquisition using airborne systems is expensive compared with satellite recording, as aircraft missions are generally flown for a single user and do not benefit from the synoptic view available to satellite platforms.

A new era of spaceborne remote sensing began when the Explorer VI of the United States obtained the first satellite picture of the Earth in August 1959 (European Space Agency 2014). From 1959 to 1972, Corona satellites of the United States were used for photographic surveillance. Civilian applications of satellite remote sensing began with the National Aeronautics and Space Administration's (NASA) Landsat series. Since 1972, NASA has lunched Landsat 1 (1972), Landsat 2 (1975), Landsat 3 (1978), Landsat 4 (1982), Landsat 5 (1984), Landsat 6 (1993, failed to reach orbit), Landsat 7 (1999), and Landsat 8 (2013). The multispectral scanner and return-beam vidicon were the imaging systems in the first generation of Landsat (then called ERTS-1). The second generation of Landsat (Landsats 4 and 5) includes an MSS imaging system and a new sensor, the Thematic Mapper. The third generation of Landsat (Landsats 6 and 7) includes an Enhanced Thematic Mapper Plus (ETM+). A review of the three decades of Landsat instruments (Landsat 1–7) can be found in the works of Mika (1997). Landsat 8 launched on February 11, 2013 and was developed as a collaboration between NASA and the U.S. Geological Survey. With two science instruments— the Operational Land Imager (OLI) and the Thermal Infrared Sensor (TIRS)— Landsat 8 represents an evolutionary advance in technology. OLI provides two new spectral bands, one for detecting cirrus clouds and the other for coastal zone observation. TIRS has two narrow spectral bands in the thermal region which was formerly covered by one wide spectral band on Landsats 4–7. The Landsat satellite series are a great contribution to remote sensing. In fact, it is the MSS aboard the first Landsat that gives most earth scientists their first look at images taken in a spectral region beyond that seen by human eyes. Other multispectral remote sensing satellites launched during this period include the Système Probatoire d'Observation de la Terre (SPOT) series developed by France—SPOT-1 (1986), SPOT-2 (1990), SPOT 3 (1993), SPOT4 (1998), SPOT 5 (2002), and SPOT 6 (2012); the India Remote Sensing Satellite (IRS) series started in 1988; the China-Brazil Earth Resource Satellite series started in 1999; and high-resolution satellites such as IKONOS (1999), QuickBird (2001), WorldView-1 (2007), GeoEye-1 (2008), WorldView-2 (2009), and China's Gaofen (high resolution) satellite series started in 2013, among others. Multispectral data collected by these spaceborne platforms have been widely used in many application fields. Figure 1.3 is a color composite of Landsat-5 TM bands 4 (red), 3 (green), and 2 (blue) acquired on August 16, 1992, near Kunming, Yunnan, China.

1.3   HYPERSPECTRAL REMOTE SENSING

Compared with multispectral remote sensing that uses relatively broad spectral bands, hyperspectral remote sensing uses imaging spectrometers that measure near-laboratory-quality spectra in narrow spectral bands. Therefore, a complete reflectance spectrum can be derived from the spectral bands for every pixel in the

FIGURE 1.3 Color composite of Landsat-5 TM bands 4 (red), 3 (green), and 2 (blue) acquired on August 16, 1992, near Kunming, Yunnan, China.

scene (Figure 1.4). It should be noted that there is no absolute threshold on the number of bands that distinguish between multispectral and hyperspectral remote sensing.

The Airborne Imaging Spectrometer (AIS) was the first of the high-resolution imaging spectrometers (Goetz et al. 1985a). The success of the AIS gave impetus to the development of an improved optical sensor, the Airborne Visible and Infrared Imaging Spectrometer (AVIRIS), which delivers calibrated images of 224 contiguous spectral channels within the wavelengths ranging from 400 to 2450 nanometers (nm)

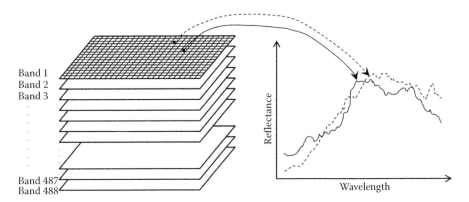

FIGURE 1.4 The imaging spectrometry concept. A spectral curve can be extracted from hundreds of spectral bands for each pixel location.

TABLE 1.1

Configuration of Five Airborne Imaging Spectrometers

Sensor	Spectral Range (nm)	Bands	Spectral Resolution (nm)	Country	First Operation
CASI	430–870	288	2.9	Canada	1989
SFSI	1200–2400	122	10.0	Canada	1993
AIS-1	900–2100 1200–2400	128	9.3	USA	1982
AIS-2	800–1600 1200–2400	128	10.6	USA	1985
AVIRIS	400–2450	224	9.4–16.0	USA	1987

(Vane and Goetz 1993). Table 1.1 lists the configuration of five early airborne imaging spectrometers.

The launch of NASA's Earth Observing-1 (EO-1) platform in November 2000 marks the first operational test of NASA's "New Millennium" spaceborne hyperspectral technology for Earth observation. The theme of the EO-1 mission is the evaluation of advanced earth observation instruments through a combination of direct measurements of performance and a broad range of application studies (Ungar et al. 2003). The Hyperion imaging spectrometer onboard the EO-1 is the first high spatial resolution imaging spectrometer to orbit the Earth. Hyperion is capable of resolving 220 spectral bands (from 0.4 to 2.5 μm) with a 30-m resolution, covering a 7.5 km × 100 km land area per image. More information on the EO-1 and Hyperion can be found in the works of Pearlman et al. (2003) and Ungar et al. (2003).

By collecting as many as hundreds of contiguous, inherently registered spectral images of the scene, the imaging spectrometers make it possible for direct identification of surface materials based on their diagnostic spectral characteristics and present the results as images, which greatly improves the discrimination of ground features and phenomena. For example, AVIRIS images have been used to measure and identify the constituents of rock units based on molecular absorption and particle scattering signatures. Applications of hyperspectral remote sensing include geologic mapping and mineral exploration (Goetz 1984, Goetz et al. 1985b, Mustard and Pieters 1987, Farrand and Singer 1991, Kruse et al. 1993, Farrand and Harsanyi 1995, Cloutis 1996) and vegetation mapping (Galvâo et al. 2005, Li et al. 2005, Dong 2008, Kalacska and Sanchez-Azofeifa 2008, Thenkabail et al. 2011), among others. While new hyperspectral sensors are being developed, a new trend in research and development is to combine multiple types of sensors on a single platform to better use the complimentary features of the sensors. For example, a LiDAR, Hyperspectral and Thermal airborne imager was developed by NASA Goddard Space Flight Center (Cook et al. 2013). Figure 1.5 shows a color composite image created from CASI-1500 Visible and Near Infrared (VNIR) hyperspectral bands acquired in Surrey, BC, Canada, in May 2013.

FIGURE 1.5 Color composite image created from CASI-1500 VNIR hyperspectral bands acquired in Surrey, BC, Canada, in May 2013. Band 70 (1.0273 μm) is used as red, band 15 (0.5020 μm) as green, and band 5 (0.4061 μm) as blue.

1.4 RADAR REMOTE SENSING

Both airborne and spaceborne passive electro-optical sensors are hindered by cloud cover and sun illumination conditions. The limitation can be removed by active radar imaging systems that operate independently of lighting conditions and largely independently of weather. A radar remote sensing system uses its own electromagnetic energy in microwave bands to "illuminate" the terrain and detects the energy returning from the terrain, with the transmitter and the receiver in the same location. The way electromagnetic waves propagate through a material can be described by a radar equation. Neglecting the path losses, the radar equation may be written as follows (Fung and Ulaby 1983):

$$P_r = \frac{P_t G_t}{4\pi R^2} \sigma_{rt} \frac{A_r}{4\pi R^2} \tag{1.1}$$

where P_r is the received power at polarization r, P_t is the transmitted power at polarization t, G_t is the gain of the transmitting antenna in the direction of the target at polarization t, R is the distance between radar and target, σ_{rt} is the radar cross section, the area intercepting that amount of incident power of polarization t which, when scattered isotropically, produces an echo at polarization r equal to that observed from the target, A_r is the effective receiving area of the receiving antenna aperture at polarization r.

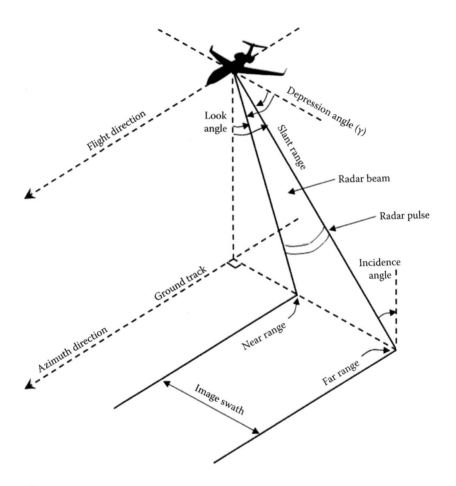

FIGURE 1.6   Concepts in airborne radar remote sensing.

The equation shows that the characteristics of radar images depend on radar system parameters, such as incidence angle, wavelength (or frequency), and polarization, and target parameters, such as complex dielectric constant, surface roughness, and volume scattering (Fung and Ulaby 1983). It is therefore important to understand how radar waves interact with natural surfaces in order to conduct correct interpretation of radar images. Figure 1.6 shows some major parameters in airborne radar remote sensing. More details of the parameters are described below.

1. Incidence angle

   The incidence angle is the angle between a radar beam and a line perpendicular to the surface. It is known as "local incidence angle" when the surface is not horizontal. Here, "depression angle (γ)" is used to refer to the angle between a horizontal plane and a beam from the antenna to a target on the ground (Figure 1.6). The character of depression angle can cause shadows on radar images. In all cases except when the depression angle is equal to 0°, all slopes facing the radar (the foreslope) are shortened relative to their true length, with most of the shortening closest to the radar. The phenomenon is referred to as "radar foreshortening" (Simonett and Davis 1983). If the curvature of a transmitted radar pulse causes the top of a tall vertical target to reflect energy in advance of its base, the distortion on the image is called radar layover. It occurs where the top of an object is closer to the radar than the bottom and is therefore recorded sooner. In mountainous areas, radar layover is often a severe problem that causes foreslopes of an image uninterpretable.

2. Dielectric properties

The relative complex dielectric constant of a material, ε, consists of a real part, ε', and an imaginary part, ε''. ε can be expressed as (Fung and Ulaby 1983)

$$\varepsilon = \varepsilon' - j\varepsilon''. \tag{1.2}$$

where ε is called the dielectric constant of the material, ε' is referred to as the relative permittivity, and ε'' as the loss factor. For terrestrial rocks in an arid environment, the density of a rock is the major factor affecting the dielectric constant, but the dielectric constant variations do not have substantial influence on radar signal because most natural rocks have dielectric constants in a narrow range (Farr 1993). For soils, the complex dielectric constant is a function of frequency, soil moisture constant, and soil type (Fung and Ulaby 1983).

3. Polarization

Polarization refers to the orientation of the electromagnetic vector of the transmitted radar signal. A radar wave may be transmitted with horizontal (H), vertical (V), or circular (left-hand circular—L, right-hand circular—R) polarization. Transmitting and receiving electrical field vectors in the same direction is known as like-polarization (HH, VV), whereas transmitting an electrical field vector in one direction (horizontal or vertical) and receiving electrical field vectors in the perpendicular direction is called cross-polarization (HV, VH). Because of the differences in the physical processes for the two types of signal returns, like-polarization images may be different from cross-polarization images. A radar system can have single polarization, dual polarizations, or four polarizations.

4. Wavelength, surface roughness, and penetration

The radar wavelength can affect the scattered signal by defining a surface roughness and by determining the depth of penetration (Fung and Ulaby 1983). Surface roughness is different from topographic relief in that it is determined by surface features comparable in size to the radar wavelength. For most natural surfaces, it is difficult to characterize them mathematically due to their complex geometry, but the root-mean-square height of the surface variations is an adequate approximation of surface relief (Sabins 1987). The theoretical boundary between smooth and rough surfaces for a given radar wavelength (λ) and depression angle (γ) can be defined by the

Rayleigh criterion or the modified criteria of Peake and Oliver (1971), or a more stringent Fraunhofer criterion for defining a radar-smooth surface (Ulaby et al. 1982):

$$h < \frac{\lambda}{32\cos\theta} \tag{1.3}$$

where h is the standard deviation of the height variation of the object, λ is the wavelength, and θ is the incidence angle referring to the vertical direction.

The radar wavelength also influences the depth of penetration of radar waves. The term "skin depth" is often used to define the depth below the surface at which the amplitude of the incident wave will have decreased to about 37% of its value at the surface (Fung and Ulaby 1983). A striking example of radar penetration is provided by the study of McCauley et al. (1982) in the Sahara Desert, where penetration of 1–6 m of the L-Band Shuttle Imaging Radar (SIR-A) was reported and details of ancient drainage patterns underlying the dry sand sheet were shown.

Like multispectral and hyperspectral remote sensing, radar remote sensing also uses airborne and spaceborne platforms. Airborne radar imagery was employed for geological investigations early in the 1960s (MacDonald 1969, Wing 1971). Since the 1980s, new airborne radar systems such as SAR-580 (Canada), STAR-1 (USA), and CASSAR (China) have been developed for various applications. Since 1978, NASA has launched four temporary spaceborne radar systems: Seasat, SIR-A, Shuttle Imaging Radar-B (SIR-B), and Shuttle Imaging Radar-C (SIR-C/X-SAR). Other spaceborne radar systems include Almaz-1 (1991, Soviet Union), ERS-1 (1991, European Space Agency), ERS-2 (1995, European Space Agency), JERS-1 (1992, Japan), Radarsat-1 (1995, Canada), Envisat-1 (2002, European Space Agency), ALOS/PALSAR (2005, Japan), and Radarsat-2 (2007, Canada). Applications of radar remote sensing can be found in many fields, including geologic mapping (Blom and Daily 1982, Cimino and Elachi 1982, Breed et al. 1983, Evans et al. 1986, Fielding et al. 1986, Lynne and Taylor 1986, Sabins 1987, Gaddis et al. 1989, Evans and van Zyl 1990, Evans et al. 1990, Singhroy et al. 1993, Singhroy and Saint-Jean 1999, Guo et al. 1993a,b, 1996, 1997, Moon et al. 1994, Mouginis-Mark 1995, Kruse 1997, Schaber et al. 1997, Wood et al. 1997, Mahmood et al. 1999, Schaber 1999, Pal et al. 2007), vegetation discrimination (Evans et al. 1986, Durden et al. 1989, Dobson et al. 1995, Harrell et al. 1997, Cloude and Treuhaft 1999, Sawaya et al. 2010, Evans and Costa 2013), and soil moisture evaluation (Dabrowska-Zielinska et al. 2002, Kasischke et al. 2007, van der Velde et al. 2012, Bourgeau-Chavez et al. 2013), among others. Schmullius and Evans (1997) analyzed radar frequency and polarization requirements for applications in ecology, geology, hydrology, and oceanography. Figure 1.7 presents SIR-C and Radarsat-1 images of a 15 km × 25 km area near Yuma, AZ, USA.

1.5 LiDAR REMOTE SENSING

LiDAR stands for Light Detection and Ranging, a technology that measures distances (or ranges) based on the time between transmitting and receiving laser signals. Both pulsed and continuous wave lasers can be used: pulsed lasers transmit

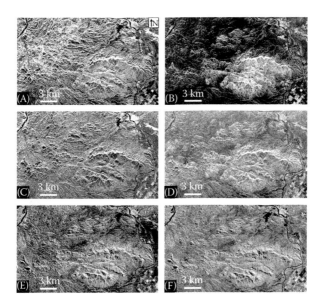

FIGURE 1.7   SIR-C and Radarsat-1 images of a 15 km × 25 km area near Yuma, AZ, USA. (A) SIR-C L-HH image; (B) SIR-C L-HV image; (C) SIR-C C-HH image; (D) SIR-C C-HV image; (E) Radarsat-1 C-HH Standard Beam 4 image; and (F) Radarsat-1 C-HH Extended High Incidence Bean 3 image.

energy of very short duration and detect ranges based on amplitudes of the received signals; in contrast, continuous wave lasers detect ranges based on the phase difference between transmitted and received signals (Baltsavias 1999b). Pulsed lasers are most often used in terrestrial applications and thus are the focus of this book.

As can be seen from Figure 1.2, LiDAR is an active remote sensing method that can be used on spaceborne, airborne, and ground-based platforms. In 2003, NASA launched the Ice Cloud and Land Elevation Satellite (ICESat) which carried the Geoscience Laser Altimeter System (GLAS), a laser profiler with 65-m footprint on the ground, to collect data about the polar ice caps, vegetation canopy, and other parameters. Airborne LiDAR is sometimes used interchangeably with Airborne Laser Scanning, Airborne Laser Swath Mapping, or Laser Radar. Ground-based LiDAR is often called Terrestrial Laser Scanning, which includes Stationary Terrestrial Laser Scanning from a static vantage point on the surface of the earth and Mobile Terrestrial Laser Scanning from a moving vehicle.

An airborne or satellite LiDAR remote sensing system typically consists of (1) a laser range finder that detects ranges and (2) a positioning and orientation system that measures the location and orientation of the sensor, which in combination can derive the three-dimensional (3D) coordinates of the objects it detects. Since LiDAR can directly measure the geographic environment in three dimensions (3D), it does not have the problem of geometric distortion (e.g., relief displacement) associated with imaging that has to project the 3D world into a two-dimensional image space. In other words, a user does not have to worry about the issue of georeferencing, a nontrivial issue for image processing. This is one of the main advantages of LiDAR.

Another advantage of LiDAR is that data can be collected at daytime or night-time, as long as there is no heavy fog, smoke, or high levels of moisture such as rain, snow, and clouds between the laser system and the object. For example, LiDAR data can be collected at night when the wind is calm (Figure 1.8).

The most useful characteristic of LiDAR might be that the laser energy can penetrate through canopy gaps and measure canopy structural and terrain elevation along the direction of laser rays. In an optical image, the value of each pixel (gray scale or color) is dominated by the reflectance of the object surface, and users cannot really see the terrain under dense canopy (Figure 1.9A). However, laser energy can reach

FIGURE 1.8 Ground-based LiDAR data collected at midnight for tree ferns in a tropical forest of HI, USA.

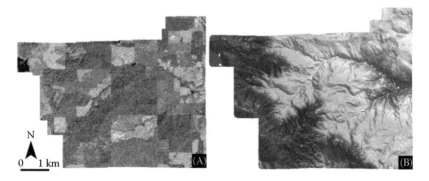

FIGURE 1.9 Different views of Panther Creek, Oregon, based on optical imagery and LiDAR data. (A) Geoeye imagery and (B) DTM derived from airborne LiDAR.

terrain so that an analyst can use the ground laser returns to generate continuous Digital Terrain Models under canopy (Figure 1.9B).

Compared with multispectral, hyperspectral, and radar remote sensing discussed in previous sections, LiDAR remote sensing is a relatively new field. Although lasers have been used for atmospheric research for decades (Goyer and Watson 1963), and laser altimeters have been used for measuring the distance from an orbiting spacecraft to the surface of the planet or asteroid, it was the development of high-accuracy global positioning systems and Inertial Measurement Units by the mid-1990s that made airborne LiDAR survey possible. The state-of-the-art LiDAR systems are capable of emitting over 1 million pulses per second. Reviews of LiDAR history and new systems can be found in the works of Baltsavias (1999a) and Mallet and Bretar (2009).

The subsequent chapters will introduce the principles of LiDAR remote sensing, LiDAR data processing, and LiDAR applications in forestry and vegetation mapping, urban environments, and geosciences. As full-waveform LiDAR data is mainly used for forest analysis and their contribution is less obvious in other application fields, the remaining chapters will focus on discrete-return LiDAR data and applications. To help readers better understand how LiDAR data is used to solve real-world problems, each of the subsequent chapters has several data processing/analysis/application projects with step-by-step instructions. Esri's ArcGIS software (version 10.2.2 or later) is used in most of the exercises, and data for the projects can be downloaded from http://geography.unt.edu/~pdong/LiDAR/. Although prior knowledge of LiDAR is not required, most of the projects assume that users have worked with vector and raster data in ArcGIS.

REFERENCES

Baltsavias, E.P., 1999a. Airborne laser scanning: Existing systems, firms, and other resources. *ISPRS Journal of Photogrammetry and Remote Sensing*, 54: 164–198.

Baltsavias, E.P., 1999b. Airborne laser scanning: Basic relations and formulas. *ISPRS Journal of Photogrammetry and Remote Sensing*, 54: 199–214.

Blom, R.G., and Daily, M., 1982. Radar image processing for rock-type discrimination. *IEEE Transaction on Geosciences and Remote Sensing*, 20: 343–351.

Bourgeau-Chavez, L.L., Leblon, B., Charbonneau, F., and Buckley, J.R., 2013. Evaluation of polarimetric Radarsat-2 SAR data for development of soil moisture retrieval algorithms over a chronosequence of black spruce boreal forests. *Remote Sensing of Environment*, 132: 71–85.

Breed, C.S., Schaber, G.G., McCauley, J.F., Grolier, M.J., Haynes, C.V., Elachi, C., Blom, R., Issawi, B., and McHugh, W.P., 1983. Subsurface geology of the western desert in Egypt and Sudan revealed by shuttle imaging radar (SIR-A). *Space Imaging Radar Symposium*, Pasadena, 17–20 January, 1983, JPL Publication 83-11, pp. 10–12.

Cimino, J.B., and Elachi, E., eds, 1982. *Shuttle Imaging Radar-A (SIR-A) Experiment*, Jet Propulsion Laboratory Publication 8277, Pasadena, CA.

Cloude, S.R., and Treuhaft, R.N., 1999. The structure of oriented vegetation from polarimetric interferometry. *IEEE Transaction on Geoscience and Remote Sensing*, 37: 2620–2624.

Cloutis, E.A., 1996. Hyperspectral geological remote-sensing—Evaluation of analytical techniques. *International Journal of Remote Sensing*, 17: 2215–2242.

Colwell, R.N., 1984. From photographic interpretation to remote sensing. *Photographic Engineering and Remote Sensing*, 50: 1305.

Cook, B.D., Corp, L.W., Nelson, R.F., Middleton, E.M., Morton, D.C., McCorkel, J.T., Masek, J.G., Ranson, K.J., Ly, V., and Montesano, P.M., 2013. NASA Goddard's Lidar, hyperspectral and thermal (G-LiHT) airborne imager. *Remote Sensing*, 5: 4045–4066, doi: 10.3390/rs5084045.

Dabrowska-Zielinska, K., Gruszczynska, M., Kowalik, W., and Stankiewicz, K., 2002. Application of multisensor data for evaluation of soil moisture. *Advances in Space Research*, 29: 45–50.

Desjardins, L., 1950. Techniques in photogeology. *Bulletin of the American Association of Petroleum Geologists*, 34: 2284–2317.

Dobson, M.C., Ulaby, F.T., Pierce, L.E., Sharik, T.L., Burgen, K.M., Kellndorfer, J., Hendra, J.R., Li, E., Lin, Y.C., Nashashibi, A., Sarabandi, K., and Siqueira, P., 1995. Estimation of forest biophysical characteristics in northern Michigan with SIR-C/X-SAR. *IEEE Transactions on Geoscience and Remote Sensing*, 33: 887–895.

Dong, P., 2008. Fractal signatures for multiscale processing of hyperspectral image data. *Advances in Space Research*, 41: 1733–1743.

Durden, S.L., van Zyl, J.J., and Zebker, H.A., 1989. Modeling and observation of the radar polarization signature of forested areas. *IEEE Transactions on Geoscience and Remote Sensing*, 27: 290–301.

European Space Agency website. http://www.esa.int/About_Us/Welcome_to_ESA/ESA_history/50_years_of_Earth_Observation (Last Accessed May 2014).

Evans, D.L., and van Zyl, J.J., 1990. Polarimetric imaging radar: Analysis tools and applications. In: (J.A. Kong, ed.) *Radar Polarimetry. Progress in Electromagnetic Research*, Volume 3, Elsevier Science Publishing Co., New York, 520 p.

Evans, T.L., and Costa, M., 2013. Landcover classification of the Lower Nhecolândia subregion of the Brazilian Pantanal Wetlands using ALOS/PALSAR, RADARSAT-2 and ENVISAT/ASAR imagery. *Remote Sensing of Environment*, 128: 118–137.

Evans, D.L., Farr, T.G., Ford, J.P., Thompson, T.W., and Werner, C.L., 1986. Multipolarization radar images for geologic mapping and vegetation discrimination. *IEEE Transactions on Geoscience and Remote Sensing*, 24: 246–257.

Evans, D.L., van Zyl, J.J., and Burnette, C.F., 1990. Incorporation of polarimetric radar images into multisensor data sets. *IEEE Transactions on Geoscience and Remote Sensing*, 28: 932–939.

Farr, T.G., 1993. Radar interactions with geologic surfaces. In *Guide to Magellan Image Interpretation*, National Aeronautics and Space Administration, Pasadena, CA, November 1, 1993.

Farrand, W.H., and Singer, R.B., 1991. Analysis of altered volcanic pyroclasts using AVIRIS data. In: *Proceedings, 3rd Airborne Visible/Infrared Imaging Spectrometer (AVIRIS) Workshop*, JPL Publication 91-28, Pasadena, CA, pp. 248–257.

Farrand, W.H., and Harsanyi, J.C., 1995. Discrimination of poorly exposed lithologies in imaging spectrometer data. *Journal of Geophysical Research-Planets*, 100: 1565–1578.

Fielding, E., Knox, W.J., Jr., and Bloom, A.L., 1986. SIR-B radar imagery of volcanic deposits in the andes. *IEEE Transactions on Geoscience and Remote Sensing*, GE-24: 582–589.

Fung, A.K., and Ulaby, F.T., 1983. Matter-energy interaction in the microwave region, In: (R.N. Colwell, ed.) *Manual of Remote Sensing* (2nd edition), Chapter 4, American Society of Photogrammetry, Falls Church, VA, Volume I, pp. 115–164.

Fussell, J., Rundquist, D., and Harrington, J.A., 1986. On defining remote sensing. *Photographic Engineering and Remote Sensing*, 50: 1507–1511.

Gaddis, L., Mouginis-Mark, P., Singer, R., and Kaupp, V., 1989. Geologic analysis of shuttle imaging radar (SIR-B) data of Kilauea Volcano, Hawaii. *Geological Society of America Bulletin*, 101: 317–332.

Galvâo, L.S., Formaggio, A.R., and Tisot, D.A., 2005. Discrimination of sugarcane varieties in Southeastern Brazil with EO-1 Hyperion data. *Remote Sensing of Environment*, 94: 523–534.

Gillespie, A.R., Kahle, A.B., and Palluconi, F.D., 1984. Mapping alluvial fans in Death Valley, California, using multichannel thermal infrared images. *Geophysical Research Letters*, 11: 1153–1156.

Goetz, A.F.H., 1984, High spectral resolution remote sensing of the land. *Proceedings, Society Photo-Optical Instrumentation Engineers*, 475: 56–68.

Goetz, A.F.H., 1989. Spectral remote sensing in geology. In: (G. Asrar, ed.) *Theory and Application of Optical Remote Sensing*, John Wiley & Sons, New York, 733 p.

Goetz, A.F.H., Wellman, J.B., and Bames, W.L., 1985a. Optical remote sensing of the Earth. *IEEE Proceedings*, 73: 950–969.

Goetz, A.F., Vane, G., Soloman, J.E., and Rock, B.N., 1985b. Imaging spectrometry for earth remote sensing. *Science*, 228: 1147–1153.

Goyer, G.G., and Watson, R., 1963. The laser and its application to meteorology. *Bulletin of the American Meteorological Society*, 44: 564–575.

Guo, H.D., Zhang, Y.H., Shao, Y., Dong, P.L., and Wang, C., 1993a. Geological analysis using shuttle imaging radar and airborne SAR in China. *Advances in Space Research*, 13: 79–82.

Guo, H.D., Shao, Y., and Dong, P., 1993b. Geological and geomorphological analysis using Shuttle Imaging Radar (SIR-B) and airborne multi-polarization data in Northern China. In: *Proceedings of the Ninth Thematic Conference on Geologic Remote Sensing*, Pasadena, CA.

Guo, H.D., Zhu, L.P., Shao, Y., and Lu, X.Q., 1996. Detection of structural and lithological features underneath a vegetation canopy using SIR-C/X-SAR data in Zhao-Qing test site of southern China. *Journal of Geophysical Research: Planets*, 101: 23101–23108.

Guo, H.D., Liao, J.J., Wang, C.L., Wang, C., Farr, T.G., and Evans, D.L., 1997. Use of multi-frequency, multipolarization shuttle imaging radar for volcano mapping in the Kunlun Mountains of Western China. *Remote Sensing of Environment*, 59: 364–374.

Harrell, P.A., Kasischke, E.S., Bourgeau-Chavez, L.L., Haney, E.M., and Christenson, N.L., Jr., 1997. Evaluation of approaches to estimating above ground biomass in southern pine forests using SIR-C data. *Remote Sensing of Environment*, 59: 223–233.

Hunt, G.R., 1980. Electromagnetic radiation: The communication link in remote sensing. In: (B.S. Siegal, and A.R. Gillespie, eds) *Remote Sensing in Geology*, John Wiley & Sons, New York, pp. 5–45.

Jensen, J.R., 2006. *Remote Sensing of the Environment: An Earth Resource Perspective* (2nd edition), Prentice Hall, Upper Saddle River, NJ, 608 p.

Kahle, A.B., 1984. Measuring spectra of arid lands. In: (F. El-Baz, ed.) *Desert and Arid Lands*: Chapter 11, Martinus Nijhoff Publishers, The Hague, Netherlands, pp. 195–217.

Kahle, A.B., and Rowan, L.C., 1980. Evaluation of multispectral middle infrared aircraft images for lithologic mapping in the East Tintic Mountains, Utah. *Geology*, 8: 234–239.

Kahle, A.B., and Goetz, A.F.H., 1983. Mineralogic information from a new airborne thermal infrared multispectral scanner. *Science*, 222: 24–27.

Kalacska, M., and Sanchez-Azofeifa, G.A., 2008. *Hyperspectral Remote Sensing of Tropical and Sub-Tropical Forests*, CRC Press, Boca Raton, FL, 352 p.

Kasischke, E.S., Bourgeau-Chavez, L.L., and Johnstone, J.F., 2007. Assessing spatial and temporal variations in surface soil moisture in fire-disturbed black spruce forests in Interior Alaska using spaceborne synthetic aperture radar imagery—Implications for post-fire tree recruitment. *Remote Sensing of Environment*, 108: 42–58.

Kruse, F.A., 1997. Comparative lithological mapping using multipolarization, multifrequency imaging radar and multispectral optical remote sensing: A SIR-C investigation. JPL SIR-C/X-SAR Final Report.

Kruse, F.A., Lefkoff, A.B., and Dietz, J.B., 1993. Expert system-based mineral mapping in northern Death Valley, California/Nevada, using the airborne visible/infrared imaging spectrometer (AVIRIS). *Remote Sensing of Environment*, 44: 309–336.

Li, L., Ustin, S.L., and Lay, M., 2005. Application of multiple endmember spectral mixture analysis (MESMA) to AVIRIS imagery for coastal salt marsh mapping: a case study in China Camp, CA, USA. *International Journal of Remote Sensing*, 26: 193–207.

Lynne, G.J., and Taylor, G.R., 1986. Geological assessment of SIR-B imagery of the Amadeus Basin, NT, Australia. *IEEE Transactions on Geoscience and Remote Sensing*, GE-24: 575–581.

MacDonald, H.C., 1969. Geologic evaluation of radar imagery from Darien Province, Panama. *Modern Geology*, 1: 1–63.

Mahmood, A., Parashar, S., and Srivastava, S., 1999. RADARSAT data applications: Radar backscatter of granitic facies, the Zaer pluton, Morocco. *Journal of Geochemical Exploration*, 66: 413–420.

Mallet, C., and Bretar, F., 2009. Full-waveform topographic lidar: State-of-the-art. *ISPRS Journal of Photogrammetry and Remote Sensing*, 64: 1–16.

McCauley, J.F., Schaber, G.G., Breed, C.S., Grolier, M.J., Haynes, C.V., Issawi, B., Elachi, C., and Blom, R., 1982. Subsurface valleys and geoarcheology of the eastern Sahara revealed by Shuttle radar. *Science*, 318: 1004–1020.

Melton, F.A., 1945. Preliminary observations on geological use of aerial photographs. *American Association of Petroleum Geologists Bulletin*, 29: 1756–1765.

Mika, A.M., 1997. Three decades of Landsat instruments. *Photogrammetric Engineering and Remote Sensing*, 63: 839–852.

Miller, V.G., 1961. *Photogeology*, McGraw-Hill, New York, 248 p.

Moon, W.M., Li, B., Singhroy, V., So, C.S., and Yamaguchi, Y., 1994. Data characteristics of JERS-1 SAR data for geological remote sensing. *Canadian Journal of Remote Sensing*, 20: 329–333.

Mouginis-Mark, P.J., 1995. Preliminary observations of volcanoes with the SIR-C radar. *IEEE Transactions on Geoscience and Remote Sensing*, 33: 934–941.

Mustard, J.F., and Pieters, C.M., 1987. Abundance and distribution of ultramafic microbreccia in Moses Rock dike: Quantitative application of mapping spectroscopy. *Journal of Geophysical Research*, 92: 10376–10390.

Pal, S.K., Majumdar, T.J., and Bhattacharya, A.K., 2007. ERS-2 SAR and IRS-1C LISS III data fusion: A PCA approach to improve remote sensing based geological interpretation. *ISPRS Journal of Photogrammetry and Remote Sensing*, 61: 281–297.

Peake, W.H., and Oliver, T.L., 1971. The response of terrestrial surfaces at microwave frequencies: Ohio State University Electroscience Laboratory, 2440-7, Technical Report AFAL-TR-70-301, Columbus, Ohio, USA.

Pearlman, J.S., Barry, P.S., Segal, C.C., Shepanski, J., Beiso, D., and Carman, S.L., 2003. Hyperion, a space-based imaging spectrometer. *IEEE Transactions on Geoscience and Remote Sensing*, 41: 1160–1172.

Sabins, F.F., Jr., 1987. *Remote Sensing Principles and Interpretation* (2nd edition), Freeman, New York, 449 p.

Sawaya, S., Haack, B., Idol, T., and Sheoran, A., 2010. Land use/cover mapping with quad-polarization radar and derived texture measures near Wad Madani, Sudan. *GIScience and Remote Sensing*, 47: 398–411.

Schaber, G.G., 1999. SAR studies in the Yuma desert, Arizona: Sand penetration, geology, and the detection of military ordnance debris. *Remote Sensing of Environment*, 67: 320–347.

Schaber, G.G., McCauley, J.F., and Breed, C.S., 1997. The use of multifrequency and polarimetric SIR-C/X-SAR data in geologic studies of Bir-Safsaf, Egypt. *Remote Sensing of Environment*, 59: 337–363.

Schmullius, C.C., and Evans, D.L., 1997. Synthetic aperture radar (SAR) frequency and polarization requirements for applications in ecology, geology, hydrology, and oceanography: A tabular status quo after SIR-C/X-SAR. *International Journal of Remote Sensing*, 18: 2713–2722.

Simonett, D.S., and Davis, R.E., 1983, Image analysis—Active microwave. In: (R.N. Colwell, ed.) *Manual of Remote Sensing* (2nd edition), Chapter 25, American Society of Photogrammetry, Falls Church, VA, Volume I, pp. 1125–1181.

Singhroy, V., and Saint-Jean, R., 1999. Effects of relief on the selection of Radarsat-1 incidence angle for geological applications. *Canadian Journal of Remote Sensing*, 25: 211–217.

Singhroy, V., Slaney, R., Lowman, P., Harris, J., and Moon, W., 1993. RADARSAT and radar geology in Canada. *Canadian Journal of Remote Sensing*, 19: 338–351.

Thenkabail, P.S., Lyon, J.G., and Huete, A., 2011. *Hyperspectral Remote Sensing of Vegetation*, CRC Press, Boca Raton, FL, 781 p.

Ulaby, F.T., More, R.K., and Fung, A.K., 1982. *Microwave remote sensing—Active and passive*, Volume II: Radar remote sensing and surface scattering and emission theory. Addison-Wesley, Reading, MA, Volume II, 608 p.

Ungar, S.G., Pearlman, J.S., Mendenhall, J.A., and Reuter, D., 2003. Overview of the earth observing one (EO-1) mission. *IEEE Transactions on Geoscience and Remote Sensing*, 41: 1149–1159.

van Nowhuys, J.J., 1937. Geological interpretation of aerial photographs. *Transactions of the American Institute of Mining Metallurgical Engineers*, 126: 607–624.

van der Velde, R., Su, Z., van Oevelen, P., Wen, J., Ma, Y., and Salama, M.S., 2012. Soil moisture mapping over the central part of the Tibetan Plateau using a series of ASAR WS images. *Remote Sensing of Environment*, 120: 175–187.

Vane, G., and Goetz, A.F.H., 1993. Terrestrial imaging spectrometry: Current status, future trends. *Remote Sensing of Environment*, 44: 117–126.

Wing, R.S., 1971. Structural analysis from radar imagery. *Modern Geology*, 2: 1–21.

Wood, C.A., Williams, S.N., Wessels, R.L., Schaefer, S., Gorman, C., England, A.W., Austin, R.T., and Hall, M.N., 1997. SIR-C radar investigations of volcanism and tectonism in the Northern Andes: Final report. JPL SIR-C/X-SAR Final Report.

2 Principles of LiDAR Remote Sensing

2.1 INTRODUCTION

Airborne LiDAR sensors can take discrete return measurements with multiple records per emitted pulse or full-waveform of a return signal at fixed time intervals such as 1 ns (about 15 cm sampling distance) (Figure 2.1). Full-waveform LiDAR is mainly used in forestry applications, whereas discrete return LiDAR can be used in many fields. This book focuses on discrete return LiDAR and applications. In this chapter, the following topics will be introduced: (1) basic components of LiDAR, (2) physical principles of LiDAR, (3) LiDAR accuracy, (4) LiDAR data formats, (5) LiDAR systems, and (6) LiDAR resources. At the end of the chapter, three projects are available for a review of zonal statistics in ArcGIS, creating a LASer (LAS) dataset and working with LiDAR data using the LAS Dataset Toolbar in ArcGIS, and visualization of LiDAR data using QT Reader (Applied Imagery) and Fugroviewer (Fugro).

2.2 BASIC COMPONENTS OF LiDAR

Lasers with a wavelength of 500–600 nm are normally used in ground-based LiDAR systems, whereas lasers with a wavelength of 1000–1600 nm are used in airborne LiDAR systems. A typical airborne LiDAR system is composed of a laser scanner; a ranging unit; control, monitoring, and recording units; differential global positioning system (DGPS); and an inertial measurement unit (IMU) (Figure 2.2). An integrated DGPS/IMU system is also called a position and orientation system that generates accurate position (longitude, latitude, and altitude) and orientation (roll, pitch, and heading) information. The laser scanning patterns can be zigzag, parallel, or elliptical (Figure 2.2). Based on ranges and scan angles, DGPS and IMU data, calibration data, and mounting parameters, collected laser points can be processed and assigned (x, y, z) coordinates in the geographic coordinate system with the World Geodetic System of 1984 (WGS84) datum (Hug 1996, Hug and Wehr 1997, Wehr and Lohr 1999).

2.3 PHYSICAL PRINCIPLES OF LiDAR

Both pulsed and continuous wave lasers are being used in LiDAR systems. Pulsed LiDAR systems measure the round-trip time of a short light pulse from the laser to the target and back to the receiver. Figure 2.3 shows amplitudes of transmitted (A_T) and received (A_R) light signals. If c is the speed of light, R is the distance between the

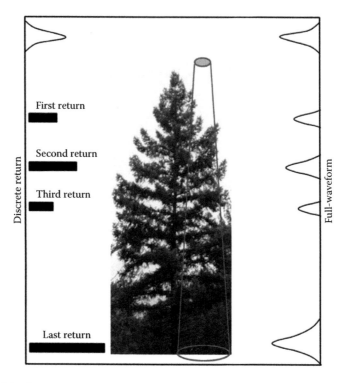

FIGURE 2.1   Discrete return and full-waveform measurement using airborne LiDAR.

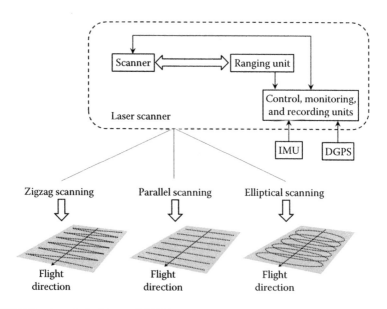

FIGURE 2.2   A typical airborne LiDAR system.

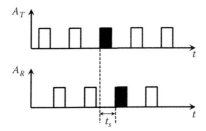

FIGURE 2.3 Amplitudes of transmitted (A_T) and received (A_R) light signals. t_s is the traveling time of a laser pulse.

ranging unit and the object surface, and t_s is the travelling time of a laser pulse; the following parameters can be calculated (Wehr and Lohr 1999):

$$\text{Range: } R = \frac{1}{2} c \cdot t_s \tag{2.1}$$

$$\text{Range resolution: } \Delta R = \frac{1}{2} c \cdot \Delta t_s \tag{2.2}$$

$$\text{Maximum range: } R_{max} = \frac{1}{2} c \cdot t_{s_{max}} \tag{2.3}$$

The LiDAR measurement process involving both detector and target characteristics is described by the standard LiDAR equation, which is derived from the radar equation (see Equation 1.1 in Chapter 1). The standard LiDAR equation relates the power of transmitted (P_t) and received (P_r) signals, and can be expressed as (Wagner et al. 2006):

$$P_r(t) = \frac{D^2}{4\pi\lambda^2} \int_0^H \frac{\eta_{sys}\eta_{atm}}{R^4} P_t\left(t - \frac{2R}{v_g}\right)\sigma(R)\,dR \tag{2.4}$$

where t is the time; D is the aperture diameter of the receiver optics; P_r is the power of received signal; P_t is the power of transmitted signal; λ is the wavelength; H is the flying height; R is the distance from the system to the target; η_{sys} and η_{atm} are the system and atmospheric transmission factors, respectively; v_g is the group velocity of the laser pulse; and $\sigma(R)\,dR$ is the apparent effective differential cross section.

The power of received signal $P_r(t)$ in Equation 2.4 can also be considered as the sum of the contribution of N targets (Mallet and Bretar 2009):

$$P_r(t) = \sum_{i=1}^{N} P_{r,i}(t) \cdot \eta_{sys}(t) \cdot \eta_{atm}(t) = \sum_{i=1}^{N} \frac{D^2}{4\pi\lambda^2 R_i^4} P_t(t) \cdot \eta_{sys}(t) \cdot \eta_{atm}(t) \cdot \sigma_i'(t) \tag{2.5}$$

where $\sigma_i'(t)$ is the apparent cross section of illuminated areas within each range interval, $P_t(t) \cdot \eta_{sys}(t)$ is the component from system contribution, and $\eta_{atm}(t) \cdot \sigma_i'(t)$ is the environment contribution.

2.4   LiDAR ACCURACY

LiDAR accuracy is usually determined by statistical comparison between known (surveyed) points and measured laser points, and is typically measured as the standard deviation (σ^2) and root mean square error (RMSE) (Evans et al. 2009). Evans et al. (2009) also suggested that methodologies for determining and reporting vertical and horizontal accuracy of discrete return LiDAR data should follow standards as outlined in FGDC-STD-007 (Federal Geographic Data Committee 1998) and NGS-58 (NOAA 1997). For bare earth surface on low to moderate slopes, the LiDAR data should conform to a minimum accuracy standard of less than 15 cm vertical and 55 cm horizontal RMSE.

The main sources of LiDAR measurement errors include those that are laser induced, problems with the inertial navigation unit (INU) for estimating positions between GPS fixes, problems with the IMU for monitoring the pointing direction of the laser, filtering induced, and other problems (Huising and Pereira 1998, Hodgson and Bresnahan 2004). Laser induced errors are normally caused by grain noise and changes in height for the points on the terrain surface at a narrow angle (such as ridges and ditches). GPS/INU/IMU errors can be caused by initialization errors and variances in the measurements. Filtering errors are related to incomplete or excessive removal of laser points. In addition, false readings from some ground features such as water bodies can also cause LiDAR measurement errors (Huising and Pereira 1998). Vertical accuracies of better than 15 cm can be obtained when the sensor altitude is below 1200 m, and up to 25 cm when the sensor altitude is between 1200 and 2500 m (Brinkman and O'Neill 2000). More detailed studies on LiDAR measurement errors can be found in the works of Schenk (2001), Ahokas et al. (2003), and Hodgson and Bresnahan (2004).

2.5   LiDAR DATA FORMATS

In the early days of LiDAR data collection, many companies used a generic American Standard Code for Information Interchange (ASCII) file interchange system. The ASCII interchange files have several major problems: (1) reading and interpreting ASCII files can be very slow, even for small amounts of LiDAR data, (2) much of the useful information is lost, and (3) the format is not standard (Figure 2.4).

For better exchange of LiDAR point cloud data, the American Society for Photogrammetry and Remote Sensing (ASPRS) introduced a sequential binary LASer (LAS) file format to contain LiDAR or other point cloud data records. The ASPRS LAS 1.0 Format Standard was released on May 9, 2003; LAS 1.1 on May 7, 2005; LAS 1.2 on September 2, 2008; LAS 1.3 on October 24, 2010; and LAS 1.4 on November 14, 2011. With LAS 1.4, the LiDAR mapping community has the ability to customize the LAS file format to meet their specific needs. The LiDAR point cloud files used in the projects of this book are all in LAS format. The specifications of all LAS versions can be accessed on the website of ASPRS.

Each LAS file could consist of a public header block, any number of Variable Length Records (VLRs), point data records, and any number of Extended Variable Length Records (EVLRs). The public header block stores basic summary information

```
99 358289.210 5973161.180 959.530 24      597546.5670 1770.970 -2482.530 182.670 188.0
99 358290.870 5973162.460 959.290 36      597546.5670 1772.030 -2482.500 182.660 173.0
5  358288.690 5973160.120 978.120 9       597546.5670 1773.080 -2482.490 182.610 188.0
99 358292.470 5973163.670 959.390 19      597546.5870 1774.240 -2483.750 182.560 182.0
5  358290.730 5973161.750 975.930 8       597546.5870 1773.100 -2483.770 182.710 150.0
5  358291.210 5973162.200 973.580 7       597546.5870 1771.930 -2483.810 182.630 176.0
5  358292.310 5973162.940 976.090 9       597546.5870 1770.870 -2483.830 182.630 176.0
5  358293.780 5973162.420 973.610 11      597546.5870 1769.720 -2483.860 182.680 183.0
5  358292.130 5973161.150 973.850 1       597546.5870 1768.560 -2483.880 182.720 172.0
5  358293.180 5973162.130 968.680 9       597546.5870 1767.410 -2483.910 182.810 171.0
5  358291.310 5973160.660 969.960 6       597546.5870 1766.350 -2483.930 182.820 176.0
10 358293.510 5973162.720 959.090 46      597546.5870 1765.200 -2483.960 182.860 195.0
10 358291.870 5973161.450 959.230 37      597546.5870 1763.950 -2483.990 182.930 182.0
5  358289.800 5973159.790 961.540 17      597546.5870 1762.890 -2484.010 182.990 158.0
10 358290.250 5973160.210 959.270 28      597546.5870 1761.740 -2484.040 183.090 166.0
```

FIGURE 2.4 Examples of LiDAR data in ASCII files. The numbers in each row are: (Left) classification code, x, y, z, and intensity; (Right) GPS time, x, y, z, and intensity.

such as the number and boundary of the points. The VLRs store information such as map projection and other metadata. The EVLRs are mainly used to store waveform data. EVLRs are specified only in LAS 1.3 and 1.4, not in the earlier versions. Therefore, waveform LiDAR data has to be stored in LAS 1.3 or higher. While the public header block and point data records are always required, the VLRs and EVLRs are optional. Therefore, an LAS file does not necessarily include key information such as the map projection of the LiDAR data. In such a case, a user needs to obtain such information from metadata files, data reports, or ask the data provider. Table 2.1 summarizes the basic structure of an LAS file.

Each record for point data stores information such as the point's x, y, z, intensity, return number, number of returns (of a given pulse), scan direction, classification, GPS time, point source, etc. (Table 2.2). Note that the number of returns indicates how many returns were received for a given transmitting pulse whereas return number indicates whether a point is the first, second,..., or last return of the pulse. For example, if a point has a value of 4 for the number of returns and a value of 2 for the return number, this means that the point is the second return of a pulse that generated four returns.

If a digital camera is integrated with a LiDAR system, each laser point can be linked with an image pixel based on photogrammetric techniques. In such a case, a point data record could also store the spectral (e.g., blue, green, red, and near-infrared) values of the associated pixel. Such spectral information is very useful for realistically visualizing the scanned landscapes in three-dimensions (3D) (Figure 2.5).

TABLE 2.1
The Basic Structure of an LAS File

LAS File Section	Note
Public header block	Required
Variable length records (VLRs)	Optional
Point data records	Required
Extended variable length records (EVLRs)	Optional

TABLE 2.2

An Example Format for LiDAR Point Data

x

y

z

Intensity

Return Number

Number of returns

Scan direction flag

Edge of flight line

Classification

Scan angle rank

User data

Point source ID

GPS time

Red

Green

Blue

Wave packet descriptor index

Byte offset to waveform data

Waveform packet size in bytes

Return point waveform location

$x(t)$

$y(t)$

$z(t)$

FIGURE 2.5 Laser points rendered based on their (left) Z elevation and (right) camera pixels' RGB spectral values.

Because of the large variety of airborne LiDAR systems (some integrate cameras and waveform digitizers while others might not) and the collected data, many point data record formats have been defined, each of which has a different record length. For example, LAS 1.0 specifies two point data record formats whereas LAS 1.4 has expanded to 10 different formats, which allows a LiDAR vendor to choose a format that can record all of the information with minimal storage requirements.

Despite the diversity of point data record formats for storing laser points in an LAS file, one key attribute that each point data record must have is the classification of the point, which tells whether the laser point is returned from the ground, vegetation, building, water, etc. (Table 2.3). Table 2.3 is an example of a standard classification scheme that is defined for point data record formats 6–10 in LAS 1.4.

For a laser point, class is probably the second most important information, next to the X, Y, and Z coordinates of the point. If laser points are not classified (i.e., the class value is 0 or 1), a LiDAR dataset is largely limited to 3D visualization of the point clouds. In contrast, a classified point cloud allows an analyst to conduct many useful analyses such as digital terrain model (DTM) generation using points classified as bare earth (class value of 2), vegetation mapping using points with class values of 3–5, and building footprint extraction using points with class value of 6.

TABLE 2.3
ASPRS Standard Classes for Point Data Record Formats 6–10 in LAS 1.4

Classification Value	Meaning
0	Created, never classified
1	Unclassified
2	Ground
3	Low vegetation
4	Medium vegetation
5	High vegetation
6	Building
7	Low point (noise)
8	Reserved
9	Water
10	Rail
11	Road surface
12	Reserved
13	Wire—guard (Shield)
14	Wire—conductor (Phase)
15	Transmission tower
16	Wire—structure connector (e.g., Insulator)
17	Bridge deck
18	Nigh noise
19–63	Reserved
64–255	User definable

A few important things should be noted to properly understand and use point classes: (1) more than one classification standards have been specified in the LAS formats. In the LAS 1.4 format, the classification standard for point data formats 0–5 is slightly different from the one for point data formats 6–10; for example, class 12 refers to "overlap points" for point data formats 0–5 and is "reserved" for formats 6–10. (2) Not every LiDAR vendor assigns the classification code in the same way. For example, points returned from water surface might be assigned to class 9 by one vendor and to class 2 by another vendor; vegetation points might be assigned to 3, 4, or 5 based on their heights by one vendor and to only 3 by another vendor. (3) Inconsistency could even exist in one LiDAR project. For example, a laser point from a building might be assigned to class 6 sometimes and to class 1 (unclassified) other times, probably by different technicians working in the project. Because of the above issues, a user should carefully check the documents provided by the LiDAR vendor to understand how points have been classified specifically. Additionally, visualization software should be used to check the accuracy of point classification (see Project 2.3).

Last but not least, compressed LiDAR binary formats have been proposed by individual developers (e.g., the .laz format) or companies (e.g., the ESRI's.zlas format). These formats can reduce the file size to ~10%–20% of the corresponding LAS files. However, they have not been endorsed by professional societies such as ASPRS.

2.6 LiDAR SYSTEMS

The National Aeronautics and Space Administration (NASA) had several experimental laser mapping systems, including the Scanning Lidar Imager of Canopies by Echo Recovery, Shuttle Laser Altimeter, Laser Vegetation Imaging Sensor, Multi-Beam Laser Altimeter, and Geoscience Laser Altimeter System. In 2013, NASA developed the Goddard's LiDAR, Hyperspectral and Thermal airborne imager (G-LiHT) for simultaneous measurements of vegetation structure, foliar spectra, and surface temperatures at very high spatial resolution (~1 m) (Cook et al. 2013). Manufactures of commercial LiDAR systems include Riegl (Austria), Toposys (Germany), TopEye/Blom (Sweden), and Optech (Canada), among others. A summary of the specifications of the NASA experimental systems and some commercial LiDAR systems can be found in the work of Mallet and Bretar (2009).

2.7 LiDAR RESOURCES

An incomplete list of LiDAR data sources and free software is provided below.
LiDAR data sources:

1. Open Topography http://www.opentopography.org
2. USGS Earth Explorer http://earthexplorer.usgs.gov
3. United States Interagency Elevation Inventory https://coast.noaa.gov/inventory/
4. National Oceanic and Atmospheric Administration (NOAA) Digital Coast https://www.coast.noaa.gov/dataviewer/#

5. Wikipedia LiDAR https://en.wikipedia.org/wiki/National_Lidar_Dataset_ (United_States)
6. LiDAR Online http://www.lidar-online.com
7. National Ecological Observatory Network—NEON http://www. neonscience.org/data-resources/get-data/airborne-data
8. LiDAR Data for Northern Spain http://b5m.gipuzkoa.net/url5000/en/G_ 22485/PUBLI&consulta=HAZLIDAR
9. LiDAR Data for the United Kingdom http://catalogue.ceda.ac.uk/ list/?return_obj=ob&id=8049, 8042, 8051, 8053

Free LiDAR software:

1. BCAL LiDAR Tools (Open source tools for visualization, processing, and analysis of LiDAR data. Requires ENVI.) http://bcal.geology.isu.edu/ Envitools.shtml
2. FugroViewer (for LiDAR and other raster/vector data) http://www. fugroviewer.com/
3. FUSION/LDV (LiDAR data visualization, conversion, and analysis) http:// forsys.cfr.washington.edu/fusion/fusionlatest.html
4. LAS Tools (Code and software for reading and writing LAS files) http:// www.cs.unc.edu/~isenburg/lastools/
5. LASUtility (A set of GUI utilities for visualization and conversion of LAS files) http://home.iitk.ac.in/~blohani/LASUtility/LASUtility.html
6. LibLAS (C/C++ library for reading/writing LAS format) http://www.liblas. org/
7. MCC-LiDAR (Multi-scale curvature classification for LiDAR) http:// sourceforge.net/projects/mcclidar/
8. MARS FreeView (3D visualization of LiDAR data) http://www.merrick. com/Geospatial/Software-Products/MARS-Software
9. Full Analyze (Open source software for processing and visualizing LiDAR point clouds and waveforms) http://fullanalyze.sourceforge.net/
10. Point Cloud Magic (A set of software tools for LiDAR point cloud visualization, editing, filtering, 3D building modeling, and statistical analysis in forestry/ vegetation applications. Contact Dr. Cheng Wang at wangcheng@radi.ac.cn)
11. Quick Terrain Reader (Visualization of LiDAR point clouds) http:// appliedimagery.com/download/

Additional LiDAR software tools can be found from the Open Topography Tool Registry webpage at http://opentopo.sdsc.edu/tools/listTools.

PROJECT 2.1:   REVIEW OF ZONAL STATISTICS FOR RASTER DATA IN ARCGIS

1. Introduction
   Zonal statistics are very useful in many applications, and will be used in Project 5-2. Zonal statistics tools in ArcGIS require two input datasets:

TABLE 2.4

Measures of Zonal Statistics in ArcGIS

Statistic	Zone Dataset	Input Value Raster	Output
Majority	Raster or polygon	Integer	Integer
Maximum	Raster or polygon	Integer, float	Same as input
Mean	Raster or polygon	Integer, float	Float
Median	Raster or polygon	Integer	Integer
Minimum	Raster or polygon	Integer, float	Same as input
Minority	Raster or polygon	Integer	Integer
Range	Raster or polygon	Integer, float	Same as input
Standard deviation	Raster or polygon	Integer, float	Float
Sum	Raster or polygon	Integer, float	Float
Variety	Raster or polygon	Integer	Integer

(1) a zone dataset which can be defined by raster cells that have the same value (such as land cover/land use types) or a polygon feature class (such as census blocks, parcels, and building footprints), and (2) a value raster. The output from zonal statistics can be a raster or a table. Table 2.4 lists 10 zonal statistical measures, along with the data types for input value rasters and output rasters or numbers. Definitions of these 10 statistical measures can be found in ArcGIS Help documents.

This project is a review of zonal statistics for raster data in ArcGIS. The data used in this project includes a polygon shapefile (parcels.shp) and six other files with different extensions for 52 parcels and a raster (bldg_heights. tif) with 0.5 m × 0.5 m cell size for heights of 52 buildings derived from LiDAR data (Figure 2.6). A perspective view of the buildings is shown in Figure 2.7. In addition to statistical measures of raster data in each parcel, a method for calculating the volume of individual buildings is demonstrated. Data for this project can be downloaded from the following webpage by right-clicking each file and saving it to a local folder: http://geography.unt. edu/~pdong/LiDAR/Chapter2/Project2.1/.

The process of calculating building volume based on zonal sum is described here. Figure 2.8 shows a building height raster and a parcel polygon (zone). Inside the parcel polygon, cells for empty space have a height value of 0, whereas building cells have positive heights.

Denote c as the cell size of the building height raster, and h_1, h_2, h_3,..., h_n as the heights of n building cells; the total volume of the building, V, in the parcel can be calculated as the sum of the volume of individual building columns:

$$
\begin{aligned}
V &= V_1 + V_2 + V_3 + \cdots + V_n \\
&= c^2 \cdot h_1 + c^2 \cdot h_2 + c^2 \cdot h_3 + \cdots + c^2 \cdot h_n \\
&= c^2 \cdot \left(h_1 + h_2 + h_3 + \cdots + h_n \right)
\end{aligned}
\tag{2.6}
$$

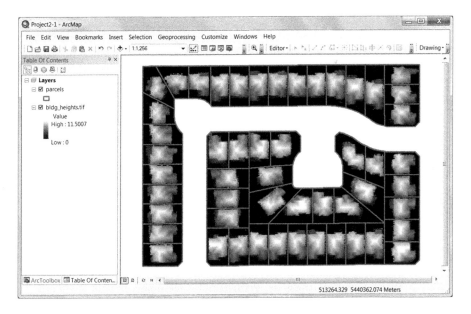

FIGURE 2.6   Parcels and height raster for 52 buildings (cell size: $0.5\,\text{m} \times 0.5\,\text{m}$).

FIGURE 2.7   Perspective view of 52 buildings derived from LiDAR data.

Since $h_1 + h_2 + h_3 + \cdots + h_n$ is the zonal sum, Equation (2.6) means that building volume can be calculated from the zonal sum of a building height raster. Alternatively, Equation (2.6) can be rearranged as:

$$V = c^2 \cdot \left(h_1 + h_2 + h_3 + \cdots + h_n \right) = c^2 \cdot n \cdot \left(h_1 + h_2 + h_3 + \cdots + h_n \right) / n \quad (2.7)$$

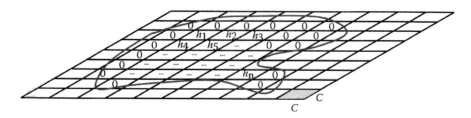

FIGURE 2.8   Building height raster and parcel polygon.

where $(h_1+h_2+h_3+\cdots+h_n)/n$ is the zonal mean. Therefore, building volume can also be calculated from the zonal mean of a building height raster. The zones can be defined by building footprints, census blocks, census tracks, or other polygons, depending on the application.

2. Project Steps
   1. Open an empty Word document so that you can copy any results from the following steps to the document. To copy the whole screen to your Word document, press the PrtSc (print screen) key on your keyboard, then open your Word document and click the "Paste" button or press Ctrl+V to paste the content into your document. To copy an active window to your Word document, press Alt+PrtSc, then paste the content into your document.
   2. Open ArcMap, and turn on the Spatial Analyst Extension.
   3. Open ArcToolbox → Spatial Analyst Tools → Zonal → Zonal Statistics as Table, use "pacels.shp" as the input feature zone data, "FID" as the zone field, "bldg_heights.tif" as the input value raster, "Zonal_Statistics" as the output table; check "Ignore NoData", select "ALL" as statistics type, and click OK to obtain the results (Figure 2.9). Note: Using the

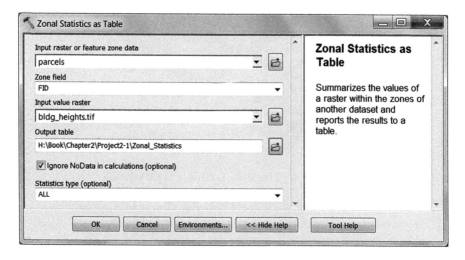

FIGURE 2.9   Zonal statistics as table tool in ArcGIS.

Rowid	FID	COUNT	AREA	MIN	MAX	RANGE	MEAN	STD	SUM	BLDG_VOLUME
1	0	1430	357.5	0	10.114115	10.114115	4.269443	3.852909	6105.30312	1526.326
2	1	1497	374.25	0	10.165668	10.165668	4.213391	3.698037	6307.445782	1576.861
3	2	1456	364	0	10.785145	10.785145	4.466122	3.907755	6502.67432	1625.669
4	3	1845	461.25	0	11.500671	11.500671	3.858065	4.171799	7118.12915	1779.532
5	4	1439	359.75	0	9.920364	9.920364	3.589593	3.829357	5165.424841	1291.356
6	5	1485	371.25	0	10.694481	10.694481	4.82669	3.613788	7167.634064	1791.909
7	6	1563	390.75	0	9.974758	9.974758	3.891149	3.598547	6081.866226	1520.467
8	7	1512	378	0	10.470276	10.470276	4.682991	3.779658	7080.682598	1770.171
9	8	1460	365	0	10.02475	10.02475	4.47508	3.512831	6533.616856	1633.404
10	9	1512	378	0	10.4016	10.4016	4.700264	3.818521	7106.79916	1776.7
11	10	1512	378	0	9.976505	9.976505	4.510839	3.595289	6820.388119	1705.097
12	11	1484	371	0	10.291059	10.291059	4.285593	3.759957	6359.820326	1589.955
13	12	1512	378	0	10.270721	10.270721	4.120725	3.910178	6230.536018	1557.634
14	13	2095	523.75	0	10.372648	10.372648	3.292861	3.730723	6898.544329	1724.636

Zonal_Statistics

FIGURE 2.10   Output table of zonal statistics with an additional field for building volume.

default output table may produce erroneous results due to unknown reasons, and it's recommended to specify an output table in a local folder.

4. The output table is shown in Figure 2.10. Since the input raster is a float raster, only six statistical measures (MIN, MAX, RANGE, MEAN, STD, and SUM) are created (see Table 2.1). It should be noted that the measurements are calculated based on parcel polygons, not just the building cells. However, as explained in Figure 2.8, the SUM field values can be used to calculate building volume. Add a new float field "BLDG_VOLUME" to the output table, then use Field Calculator and expression 0.5*0.5*[SUM] to obtain the BLDG_VOLUME field values (Figure 2.10).

5. Save your ArcMap project and Word document.

PROJECT 2.2:   CREATING AN LAS DATASET USING LiDAR POINT CLOUDS FROM FREMONT, CA, USA

1. Introduction

An LAS dataset is a stand-alone file that references one or more LiDAR data files in the LAS format. An LAS dataset can also have feature classes for surface constraints, such as breaklines, water polygons, and area boundaries. There are up to three files associated with an LAS database: LAS dataset file (.lasd), LAS auxiliary file (.lasx), and projection file (.prj). The .lasd file only stores references to actual LAS files and surface constraints, but does not import LiDAR point data from LAS files. A .lasx file is created when statistics are calculated for any LAS file in an LAS dataset. The .lasx file provides a spatial index structure that helps improve the performance of an LAS dataset. If LAS files do not have a spatial reference or have an incorrect spatial reference defined in the header of the LAS file, a projection file (.prj) can be created for each LAS file. In that case, the new coordinate system information in the .prj file will take precedence over the spatial reference in the header section of the LAS file.

In this project, an LAS dataset will be created using LiDAR data in the LAS format from Fremont, CA, USA. Since correct spatial reference is defined in the header of the LAS file, a projection file (.prj) is not needed. Once the LAS dataset is created, the user can start using the LAS dataset toolbar. Tools on the LAS dataset toolbar will also be used in this project.

2. Data

LiDAR data from a 1 km × 1 km area in Fremont, CA, USA is used in this project. The LiDAR data was collected in 2007 with a point density of 5.17 points/m^2. The horizontal coordinate system is UTM Zone 10 N WGS84 Meters [EPSG: 32610], and the vertical coordinate system is WGS84 datum. This dataset is based on services provided to the Plate Boundary Observatory (PBO) by NCALM (http://www.ncalm.org). PBO is operated by UNAVCO for EarthScope (http://www.earthscope.org) and supported by the National Science Foundation (No. EAR-0350028 and EAR-0732947). An LAS file "Fremont.las" can be downloaded (by right-clicking the file and saving it to a local folder) from: http://geography.unt.edu/~pdong/ LiDAR/Chapter2/Project2.2/

3. Project Steps

1. Open an empty Word document so that you can copy any results from the following steps to the document. To copy the whole screen to your Word document, press the PrtSc (print screen) key on your keyboard, then open your Word document and click the "Paste" button or press Ctrl+V to paste the content into your document. To copy an active window to your Word document, press Alt+PrtSc, then paste the content into your document.

2. Open ArcMap, go to the Customize menu, select "Extensions...", and turn on the 3D Analyst Extension.

3. Create an LAS dataset. Open ArcToolbox → Data Management Tools → LAS Dataset → Create LAS Dataset. Use Fremont.las as input and Fremont.lasd as output, check the "Compute Statistics" option, and then click OK to create the LAS dataset. The LAS dataset is added to ArcMap automatically. Click the "Customize" menu of ArcMap, and then select Toolbars → LAS Dataset to load the LAS dataset toolbar (Figure 2.11). Since noises (such as flying birds) are not removed from the point clouds, it appears that the point clouds have substantial variations in elevation. Note that the ground elevation of the LAS dataset is negative because the dataset has an ellipsoid-based vertical coordinate system.

4. Calculate the LAS dataset statistics. Open ArcToolbox → Data Management Tools → LAS Dataset → LAS Dataset Statistics, and use Fremont.lasd as the input LAS dataset and FremontLSD.txt as the output to generate a statistics report file (Figure 2.12). The report file is automatically added to ArcMap as a table to show statistics of the LAS dataset (Figure 2.13). For example, there are 2,411,009 ground points in the 1 km × 1 km study area, which is 45.79% of all points.

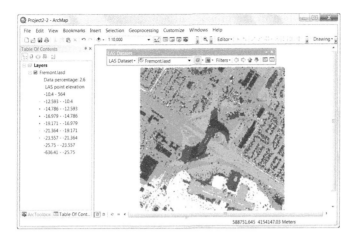

FIGURE 2.11 LAS dataset toolbar and LAS dataset of the study area.

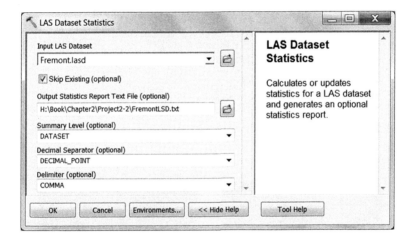

FIGURE 2.12 Calculating LAS data statistics.

Item	Category	Pt_Cnt	Percent	Z_Min	Z_Max	Intensity_Min	Intensity_Max	Synthetic_Pt_Cnt	Range_Min	Range_Max
Unknown	Returns	5265284	100	-636.41	564	<Null>	<Null>	<Null>	<Null>	<Null>
Last	Returns	5265284	100	-636.41	564	<Null>	<Null>	<Null>	<Null>	<Null>
All	Returns	5265284	100	-636.41	564	<Null>	<Null>	<Null>	<Null>	<Null>
1_Unclassified	ClassCodes	2809746	53.36	-26.5	6	0	284	0	<Null>	<Null>
2_Ground	ClassCodes	2411009	45.79	-26.54	-13.02	0	263	0	<Null>	<Null>
5_High_Vegetation	ClassCodes	9649	0.18	-636.41	564	0	230	0	<Null>	<Null>
7_Low_Point(noise)	ClassCodes	34880	0.66	-211.96	-6.98	0	71	0	<Null>	<Null>
Return_No	Attributes	<Null>	<Null>	<Null>	<Null>	<Null>	<Null>	<Null>	0	0
Intensity	Attributes	<Null>	<Null>	<Null>	<Null>	<Null>	<Null>	<Null>	0	284
Class_Code	Attributes	<Null>	<Null>	<Null>	<Null>	<Null>	<Null>	<Null>	0	7
Scan_Angle	Attributes	<Null>	<Null>	<Null>	<Null>	<Null>	<Null>	<Null>	0	0
User_Data	Attributes	<Null>	<Null>	<Null>	<Null>	<Null>	<Null>	<Null>	0	0
Point_Source	Attributes	<Null>	<Null>	<Null>	<Null>	<Null>	<Null>	<Null>	8	14
Model_Key	ClassFlags	0	0	<Null>	<Null>	<Null>	<Null>	<Null>	<Null>	<Null>
Synthetic	ClassFlags	0	0	<Null>	<Null>	<Null>	<Null>	<Null>	<Null>	<Null>
WithHeld	ClassFlags	0	0	<Null>	<Null>	<Null>	<Null>	<Null>	<Null>	<Null>

FIGURE 2.13 LAS dataset statistics report file.

The elevation of the ground surface in the study area changes from −26.54 to −13.02 m (the dataset has negative ground elevation because it has an ellipsoid-based vertical coordinate system).

5. Use LAS dataset toolbar. Explore point symbology renderers, surface symbology renderers, filter, and pan tools (Figure 2.14).
6. Create LAS dataset profile views. Use the LAS dataset profile view tool to draw a straight line over the LAS dataset, and then move the cursor to change the selection box and click on ArcMap to create a profile view. Figure 2.15 shows several profile views from the LAS dataset.

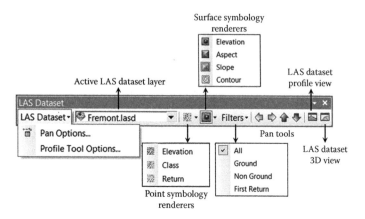

FIGURE 2.14 LAS dataset toolbar.

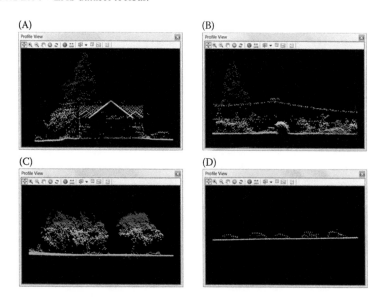

FIGURE 2.15 Sample profile views of LAS dataset for Fremont, CA, USA. (A) Roofs, trees, vehicle, and ground; (B) trees, power lines, and ground; (C) trees and ground; and (D) vehicles on road surface.

FIGURE 2.16   3D view of LAS dataset.

7. Create LAS dataset 3D views. Use the LAS dataset 3D view tool to create a 3D view, and use tools on the 3D view window to navigate the view (Figure 2.16).
8. Save your ArcMap project and Word document.

PROJECT 2.3:   EXPLORING AIRBORNE LiDAR DATA

1. Introduction
   The objective of this project is to become familiar with some common LiDAR visualization software and to improve the understanding of airborne LiDAR data. The project area is the University of Hawaii at Mānoa campus (Figure 2.17).

FIGURE 2.17   The University of Hawai'i at Mānoa campus.

The LiDAR file is named UHM.las and can be downloaded from: http://geography.unt.edu/~pdong/LiDAR/Chapter2/Project2.3/. The software tools used are Quick Terrain (QT) Reader (Applied Imagery) and Fugro. The software tools are free and can be downloaded from the vendors' websites. The following steps assume that you have installed the software on your computer.

2. Project steps
    1. Open an empty Word document so that you can copy any results from the following steps to the document. To copy the whole screen to your Word document, press the PrtSc (print screen) key on your keyboard, then open your Word document and click the "Paste" button or press Ctrl+V to paste the content into your document. To copy an active window to your Word document, press Alt+PrtSc, then paste the content into your document.
    2. Start QT Reader and open UHM.las (choose File → Open Model(s) from the menu). Visually explore the data using your mouse: scroll with the middle button to zoom in/out, hold the left button and move to rotate the point cloud, or hold the right button and move to pan the point cloud. Write down at least five types of objects (e.g., trees and buildings) you can recognize from the point cloud.
    3. Explore different display modes by clicking Display → Show/Hide. Toggle on/off "Use Height Coloration" and "Use Vertex Colors" and display the study area in three colorations similar to Figure 2.18. Explain what each coloration represents (Hint: Go to Display → Settings → Height Coloration Settings to understand the Height Coloration Setting).
    4. Go to Analysis → Model Statistics and finish the following table:

LiDAR Information	Value
Map projection	
Min/max "Return Number"	
Min/max "Number of Returns"	
Min/max "Scan Direction"	
Min/max "Line Edge"	
Min/max "Classification"	
Min/max "Scan angle"	
All "Point Source ID"	

    5. Start Fugro and open UHM.las. Click File → Lidar File Info and finish the following table:

LiDAR Information	Value
Date and year of data acquisition	
LAS version	
Number of VLRs	
Number of 1st, 2nd, 3rd, and 4th returns	

    6. Click 3D on the toolbar to open 3D view. Display the data in the following coloration settings: elevation, intensity, classification, source ID,

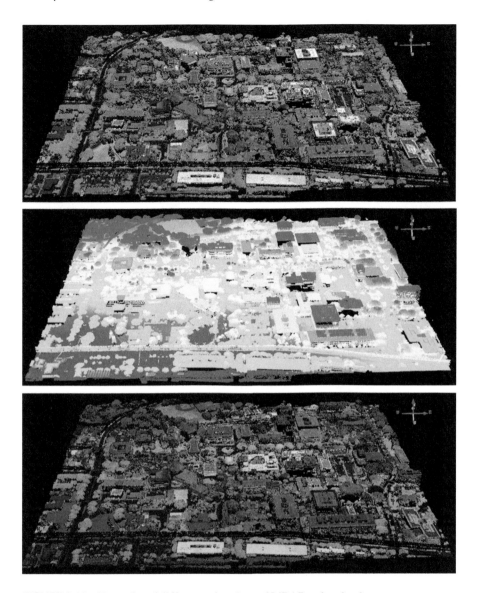

FIGURE 2.18 Examples of different colorations of LiDAR point clouds.

and return number. Take screenshots and include them in your report (see Figure 2.19 for an example). Explain what "source ID" means in this file based on what you see.

7. Pick a building on campus and use the Query Data tool 🔍 to determine its height. Document your steps of finding the height in your report, in which you should include screenshot of the building you select.

8. Pick a large tree on campus and display the point cloud with coloration based on "Return Number" (see Figure 2.20 for an example). Make sure

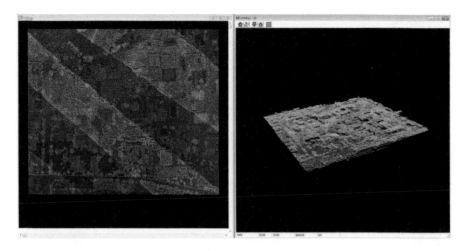

FIGURE 2.19   2D and 3D visualization of LiDAR point clouds.

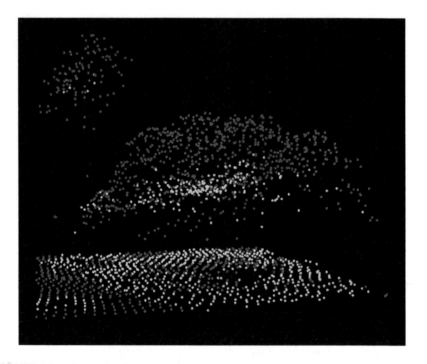

FIGURE 2.20   Example of a tree's point cloud rendered with "Return Number".

that the tree you picked includes at least three types of returns (first, second, and third returns). Identify any 3D distribution patterns you see for the different returns and explain the pattern. Include in your report a screenshot of the tree you select.

9. Save your Word document.

REFERENCES

Ahokas, E., Kaartinen, H., and Hyyppä, J., 2003. A quality assessment of airborne laser scanner data. *International Archives of Photogrammetry, Remote Sensing and Spatial Information Sciences*, 34 (Part 3/W13): 1–7.

Brinkman, R., and O'Neill, C., 2000. LiDAR and photogrammetric mapping. *The Military Engineer*, May 2000: 56–57.

Cook, B.D., Corp, L.W., Nelson, R.F., Middleton, E.M., Morton, D.C., McCorkel, J.T., Masek, J.G., Ranson, K.J., Ly, V., and Montesano, M.P., 2013. NASA Goddard's LiDAR, hyperspectral and thermal (G-LiHT) airborne imager. *Remote Sensing*, 5: 4045–4066.

Evans, J.S., Hudak, A.T., Faux, R., and Smith, A.M.S., 2009. Discrete return lidar in natural resources: Recommendations for project planning, data processing, and deliverables. *Remote Sensing*, 1: 776–794.

Federal Geographic Data Committee (FGDC), 1998. Geospatial Positioning Accuracy Standards Part 3 (FGDC-STD-007): National Standard for Spatial Data Accuracy. Reston, Virginia.

Hodgson, M., and Bresnahan, P., 2004. Accuracy of airborne lidar-derived elevation: Empirical assessment and error budget. *Photogrammetric Engineering and Remote Sensing*, 70: 331–339.

Hug, C., 1996. Combined use of laser scanner geometry and reflectance data to identify surface objects. Proceedings of the OEEPE Workshop "3-D City Models", October 9–11, 1996. Institut für Photogrammatrie, Universität Bonn.

Hug, C., and Wehr, A., 1997. Detecting and identifying topographic objects in imaging laser altimeter data. *International Archives of Photogrammetry and Remote Sensing (IAPRS)*, 32:19–26.

Huising, E., and Pereira, L.G., 1998. Errors and accuracy estimates of laser data acquired by various laser scanning systems for topographic applications. *ISPRS Journal of Photogrammetry and Remote Sensing*, 53: 1245–1261.

Mallet, C., and Bretar, F., 2009. Full-waveform topographic lidar: State-of-the-art. *ISPRS Journal of Photogrammetry and Remote Sensing*, 64: 1–16.

Schenk, T., 2001. Modeling and analysing systematic errors in airborne laser scanners. Technical report. Department of Civil and Environmental Engineering and Geodetic Science, Ohio State University, Columbus, OH.

Wagner, W., Ullrich, A., Ducic, V., Melzer, T., and Sudnicka, A., 2006. Gaussian decomposition and calibration of a novel small-footprint full-waveform digitising airborne laser scanner. *Journal of Photogrammetry and Remote Sensing*, 60: 100–112.

Wehr, A., and Lohr, U., 1999. Airborne laser scanning—An introduction and overview. *ISPRS Journal of Photogrammetry and Remote Sensing*, 54: 68–82.

3 Basics of LiDAR Data Processing

3.1 INTRODUCTION

As explained in Section 2.2 of Chapter 2, values of ranges and scan angles, differential global positioning system (DGPS) and inertial measurement unit (IMU) data, calibration data, and mounting parameters can be combined to derive (x, y, z) coordinates of LiDAR points. It is worth mentioning that since global positioning system (GPS) uses the World Geodetic System of 1984 (WGS84) datum, the originally derived three-dimensional (3D) coordinates are also georeferenced to the WGS84 datum and its ellipsoid (this means that z is ellipsoidal elevation). However, LiDAR users typically need orthometric elevation (i.e., elevation above mean sea level or a geoid model) for hydrological applications such as flooding or sea-level rise analysis. Moreover, each country or region often uses a horizontal datum that is more locally relevant (e.g., the NAD83 for the United States and Canada) in geographic information system (GIS) analysis. Therefore, a LiDAR vendor often needs to transform the original 3D coordinates of laser points from WGS84 to a new horizontal datum and/or a new vertical datum, especially upon the request of the end users.

The processing and analysis methods for LiDAR data are usually application-specific (for example, multi-scale modeling for building extraction), and many new methods are being proposed. Therefore, it is difficult to provide a complete list of various LiDAR data processing and analysis methods. However, two basic steps are usually needed: (1) classification of laser points and (2) interpolation of discrete points into a continuous surface. For example, the generation of digital terrain model (DTM) requires the classification/extraction of ground returns and the interpolation of ground returns into raster or triangulated irregular network (TIN).

The classification of laser points means identifying the type of earth surface materials or objects that generate the laser returns/points. For example, a laser pulse that is transmitted from an airplane to a tree may generate three returns: the first return from the tree crown surface, the second return from the branches and foliage below the surface, and the last return from the ground below the tree. Some users may only want to classify the points into ground versus non-ground returns because their purpose is to use the ground returns to generate a DTM. However, a forester who is interested in analyzing canopy vertical structure might want to separate the points into three categories: on canopy surface, within canopy, and on the ground. Therefore, the classification scheme used in different applications could vary substantially. Despite these, a common class that needs to be extracted is the ground returns. This is because (1) ground returns are needed to generate "elevation" models for topography (i.e., DTM), arguably the most widely used geospatial data,

41

and (2) DTM is needed to produce different "height" products for surface objects above topography. For example, building height can be calculated as the difference between elevation of laser points on a building's roof and DTM. Extracting ground returns from a point cloud is usually the first step of LiDAR data processing and the most important classification. Conventionally, this step is called "filtering" because the main driver of adopting airborne LiDAR data in its infancy stage (in the 1990s) was to generate DTM, which needs filtering out or removing non-ground returns. Because of its importance, filtering is introduced in Section 3.2 while classification of remaining non-ground returns is introduced in Section 3.3.

Unlike optical or radar imagery, airborne LiDAR data do not continuously measure or map the earth's surface. Each laser pulse and its returns are essentially samples of the environment, even at very high point density. Interpolation is needed to produce spatially continuous digital products or maps from these discrete points. For example, interpolating the ground returns and the highest returns will generate DTM and DSM (digital surface model), respectively. Although various interpolation methods exist in software, they have to be fast enough to process massive LiDAR points and intelligent enough to predict the elevation at the unsampled locations. Section 3.4 will discuss some common interpolation methods.

Figure 3.1 is a typical flowchart showing the major steps for LiDAR point data processing. After collecting (x, y, z) coordinates of LiDAR points, sorting of the points can improve the efficiency of data rendering and processing (Auer and Hinz 2007, Scheiblauer 2014, Shen et al. 2016). In addition to geometric processing of LiDAR points, it should be noted that LiDAR intensity information can also be useful. A review of LiDAR radiometric processing can be found in the work of Kashani et al. (2015). In the remainder of this chapter, we will introduce filtering, classification of non-ground returns, and spatial interpolation, respectively. Two ArcGIS projects are then presented to (1) create a DTM, a DSM, and a digital height model (DHM) for an area in Indianapolis, IN (USA), and (2) create a terrain dataset for an area in St. Albans, VT (USA).

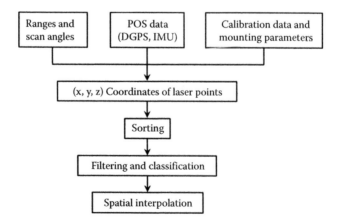

FIGURE 3.1   Flowchart of LiDAR point data processing.

3.2 FILTERING

Filtering is used to remove non-ground LiDAR points so that bare-earth digital elevation models can be created from the remaining ground LiDAR points. Over the past two decades, many filtering methods have been developed (Axelsson 2000, Sithole and Vosselman 2004, Arefi and Hahn 2005, Tóvári and Pfeifer 2005, Chen et al. 2007, Kobler et al. 2007, Liu 2008, Chang et al. 2008, Meng et al. 2010, Wang and Tseng 2010, Chen et al. 2013, Pingel et al. 2013, Zhang and Lin 2013, Lin and Zhang 2014, Zhang et al. 2016, Nie et al. 2017). Most filtering methods are unsupervised classifiers, which mean that users do not need to collect training data for ground and non-ground returns. However, they may have one or multiple parameters that need users to specify the values.

The design of a filtering algorithm is usually based on two criteria: (1) the ground has the lowest elevation compared to the objects above it (e.g., Kobler et al. 2007), and (2) elevation and slope change more slowly for bare earth than for DSM (e.g., Chen et al. 2007). However, the specific techniques that implement these criteria differ so dramatically that it is not possible to explain the details of each. As a start, readers can refer to Sithole and Vosselman (2004), which summarized characteristics of filtering algorithms from different aspects including data structure, measure of discontinuity, and filtering concepts.

Among the large variety of filtering algorithms that have been developed, surface-based approaches are probably the most popular and effective ones and have been implemented in many commercial or free software (e.g., Kraus and Pfeifer 1998, Axelsson 2000, Chen et al. 2007). Surface-based approaches usually start with an initial surface that approximates the bare earth and then generates another approximate surface of the bare earth that utilizes the information from the previous step. This process could be repeated iteratively until the next surface does not substantially differ from the previous one. This is similar to the process of k-means or ISODATA (Iterative Self-Organizing Data Analysis Technique) classification for which the algorithm stops when the classification results from the next round do not differ from the previous round (in other words, the algorithm stabilizes or converges). What is unique to LiDAR point cloud filtering is that an approximation of the bare earth surface has to be generated at each step, which is usually, but not always, implemented using interpolation methods. Surface-based approaches can be based on either TIN or raster.

3.2.1 TIN-BASED METHODS

TIN is a vector-based data structure for representing continuous surface. Specifically, a TIN surface consists of a tessellated network of non-overlapping triangles, each of which is made of irregularly distributed points. TIN is well suited for constructing terrain surface from LiDAR points because: (1) LiDAR points are often irregularly distributed due to variations in scan angle, attitude (pitch, yaw, and roll) of the airplane, and overlaps between flight lines; (2) adding or removing points into TIN can be implemented locally without reconstructing the whole TIN; and (3) the speed of constructing a TIN is usually much faster than the grid-based interpolation methods.

They are many different triangulation networks, but the Delaunay triangulation is the standard choice.

The algorithm proposed by Axelsson (2000) is probably the most famous TIN-based method, which works as follows: (1) the lowest points within a coarse grid are chosen as seed ground returns, which are used to construct an initial TIN. The grid size should be large enough (e.g., 50–100 m) to ensure that the lowest point within each grid cell are ground returns, (2) points are added into the TIN if they are close to the triangular facet and the angles to their overlaying triangular nodes are small, and (3) the densification of the TIN continues until no more ground returns can be added.

The main idea of Axelsson (2000) is to start with a small set of ground returns and then iteratively add the remaining the ground returns. Different from such an "addition" strategy, ground returns can also be extracted via "subtraction": start with all returns and then iteratively remove non-ground returns; the remaining points in the end of iteration are ground returns. For example, Haugerud and Harding (2001) first used all returns to generate a TIN surface, which was smoothed using a 3 × 3 moving window; then elevation difference was calculated between each laser return and the smoothed surface. If the difference was larger than a threshold, the point was removed. This process was repeated until a convergence threshold was reached. Note that their algorithm was designed for forest areas and it cannot efficiently remove large buildings with flat roofs. Because non-ground returns and their TIN facets appear as spikes on top of DTM, such "subtraction" algorithms are also called "despike" methods.

The TIN-based method proposed by Axelsson (2000) has some limitations in removing points belonging to lower objects and preserving ground measurements in topographically complex areas. Nie et al. (2017) proposed a revised progressive TIN densification method for filtering airborne LiDAR data using three major steps (Figure 3.2): (1) specify key input parameters; (2) select seed ground points and construct an initial Delaunay TIN; and (3) iterative densification of the TIN. Both

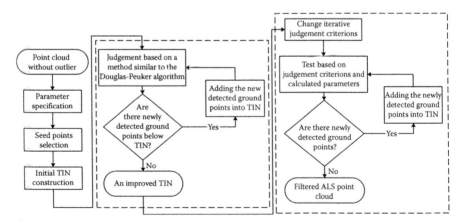

FIGURE 3.2 Flowchart of the revised progressive TIN densification method for filtering airborne LiDAR points. (Adapted from Nie, S. et al. *Measurement*, 104: 70–77, 2017.)

qualitative and quantitative analyses suggest that the revised progressive TIN densification method can produce more accurate results than the method proposed by Axelsson (2000).

3.2.2 RASTER-BASED METHODS

An alternative data structure for terrain surface is raster. Various interpolation methods (also called interpolators) can be used to generate raster grids from points. Interpolators can be exact or inexact, depending on whether the interpolated surface will go through the points or not. Interpolators can also be different depending on whether the interpolated elevations exceed the range of LiDAR point elevations. This is different from TIN because the interpolated values always go through the LiDAR points (an exact interpolator) and never exceed the range of LiDAR points. Therefore, using rasters to store terrain surface offers much flexibility for the algorithm developer to choose an appropriate interpolator. The interpolators that were often used are kriging, thin-plate spline (TPS), inverse distance weighting (IDW), and natural neighbors.

Similar to TIN-based approaches, raster-based filtering algorithms can be designed based on either an "addition" or "subtraction" strategy. As an example of the "additive" method, Chen et al. (2013) developed a multiresolution hierarchical classification (MHC) algorithm for separating ground from non-ground LiDAR point clouds based on point residuals from the interpolated raster surface. The MHC algorithm uses three levels of hierarchy from coarse to high resolutions, and the surface is iteratively interpolated towards the ground using TPS at each level, until no ground points are classified. The classified ground points are then used to update the ground surface in the next iteration.

Kraus and Pfeifer (1998) were among the first to develop a "subtractive" raster filtering algorithm. Using simple kriging (also called linear interpolator in their paper), they first fitted a surface using all returns. Then, points with large positive residuals are removed and the rest are assigned with weights according to their residuals: points with large negative residuals are more likely to be ground returns and therefore assigned with larger weights. The surfaces are iteratively refitted with remaining points with weights until convergence.

Many algorithms interpolate raster grids and filter points at multiple spatial resolutions or scales. Some progress from fine to coarse scales. For example, Evans and Hudak (2007) implemented a raster-based "subtraction" algorithm, which modified the method of Haugerud and Harding (2001) to generate raster surfaces at three gradually decreasing resolutions using TPS interpolation to iteratively remove non-ground returns. In contrast, some algorithms use gradually increasing spatial resolutions (Mongus and Žalik 2012). For example, Mongus and Žalik (2012) started with an interpolation at a spatial resolution of $64\,m \times 64\,m$ and identified non-ground returns based on the interpolated surface; the interpolation and filtering processes gradually increased to 32, 16, ..., and 1 m resolutions.

One of the main challenges in interpolation from points to raster is the demanding computation involved because (1) a LiDAR file typically has several million points per square kilometer, and (2) interpolation usually involves the use of points within each point's local neighborhood, which requires extra computation resources for indexing and searching. Therefore, an efficient algorithm would minimize the use of interpolation in its filtering process. For example, Chen et al. (2007) first created a fine-resolution (e.g., 1 m) raster grid that contains the elevation of the lowest points and then used image-based morphological operations (more specifically, opening) to remove aboveground objects such as buildings and trees. To identify buildings of different sizes, they used neighborhood windows of gradually increasing sizes for morphological opening. Their algorithm used the fact that buildings have abrupt elevation changes at their edges to separate buildings from small terrain bumps and thus is called edge-based morphological methods (Chen 2009). Ground points were identified by comparing each point with the final morphologically opened raster (which approximate the bare earth). Kriging was used finally to interpolate a smooth DTM. Since interpolation is used only once, the algorithm is fast and efficient.

3.3   CLASSIFICATION OF NON-GROUND POINTS

After separating ground and non-ground LiDAR points through filtering, non-ground points can be further classified into buildings (Axelsson 1999), roads (Choi et al. 2008), vegetation (Cobby et al. 2003), and other classes. Various methods have been proposed for LiDAR point classification, including unsupervised classification (Haala and Brenner 1999, Vosselman 2000), bayesian networks (Stassopoulou and Caelli 2000), decision trees (Antonarakis et al. 2008), and support vector machines (Charaniya et al. 2004, Lodha et al. 2006, Secord and Zakhor 2007, Mallet et al. 2011, Lin et al. 2014). In addition, LiDAR intensity data (Flood 2001, Hui et al. 2008, Niemeyer et al. 2014) and multispectral image data (Bork and Su 2007, Secord and Zakhor 2007) can be combined with LiDAR point data for classification. A review of some of the LiDAR point classification methods can be found in the work of Yan et al. (2015). Figure 3.3 is a classification map of LiDAR points, and Figure 3.4 is a profile of LiDAR points derived from P-P′ in Figure 3.3.

3.4   SPATIAL INTERPOLATION

As discussed in Section 3.2, spatial interpolation is often used in the middle of the filtering process to identify ground points. No matter which filtering method is used, interpolation is needed in the end to generate a DTM from the filtered ground returns. Some additional LiDAR-based analysis such as building footprint analysis or tree mapping (introduced in later chapters) are also needed to generate other continuous surface models such as DSM or DHM. Therefore, interpolation is a key technique in LiDAR data processing and analysis.

Spatial interpolation can be defined as predicting the values of a primary variable (such as elevation, temperature, etc.) at point locations within the same region of sampled locations. Li and Heap (2008) described a total of 38 methods in three categories (non-geostatistical, geostatistical, and combined) for spatial

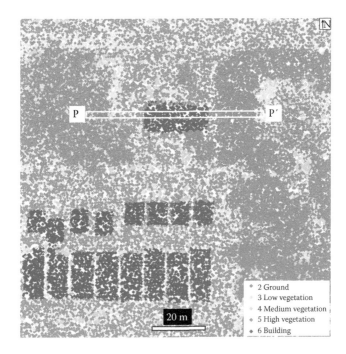

FIGURE 3.3 Classification of LiDAR points. P-P′ is the location of a profile shown in Figure 3.4.

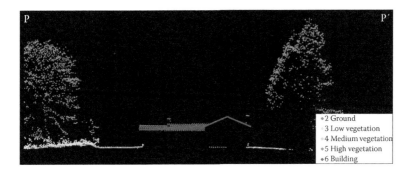

FIGURE 3.4 Profile of LiDAR points derived from P-P′ in Figure 3.3.

interpolation. A comprehensive review of spatial interpolation in the environmental sciences can be found in the work of Li and Heap (2014). Additional information on spatial interpolation in geospatial analysis is available from de Smith et al. (2015). For LiDAR point data, spatial interpolation is normally used to create DTMs from ground points, and DSMs from the highest points within cells. This section briefly introduces two common spatial interpolation methods for LiDAR point data: IDW and natural neighbor.

IDW interpolation explicitly relies on Tobler's First Law of Geography: "Everything is related to everything else, but near things are more related than distant things." (Tobler 1970). A simple model of IDW interpolation can be expressed as:

$$Z_j = \begin{cases} \dfrac{\sum_{i=1}^{n}\left(\dfrac{Z_i}{d_i^p}\right)}{\sum_{i=1}^{n}\left(\dfrac{1}{d_i^p}\right)}, & \text{if } d_i \neq 0 \text{ for all } i \\[4mm] Z_i & \text{if } d_i = 0 \text{ for some } i \end{cases} \tag{3.1}$$

where Z_i is the observation at the ith point, Z_j is the interpolated value at output location j, d_i is the distance between the ith input point and the output location j, and p is the power of distance. If $d_i = 0$ for some i, the observation Z_i is used as the output. If $p = 1$, a simple linear distance is used. A faster rate of distance decay may be obtained if $p > 1$, but a common practice is to use $p = 1$ or $p = 2$. Figure 3.5 shows how to calculate the interpolated value Z_i at the output cell location (triangle) from four observations (z_1, z_2, z_3, and z_4) using IDW interpolation with $p = 2$. More discussions on IDW interpolation can be found in the work of de Smith et al. (2015).

Natural neighbor is an interpolation method developed by Sibson (1981). Natural neighbor interpolation uses weights for each of the input points based on their area of influence, which is determined by Voronoi (Thiessen) polygons around each input point (Figure 3.6). First, a Voronoi diagram is constructed for all points (Figure 3.6A).

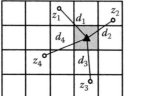

$$Z_j = \dfrac{\dfrac{z_1}{d_1^2} + \dfrac{z_2}{d_2^2} + \dfrac{z_3}{d_3^2} + \dfrac{z_4}{d_4^2}}{\dfrac{1}{d_1^2} + \dfrac{1}{d_2^2} + \dfrac{1}{d_3^2} + \dfrac{1}{d_4^2}}$$

FIGURE 3.5 Inverse distance weighting (IDW) interpolation.

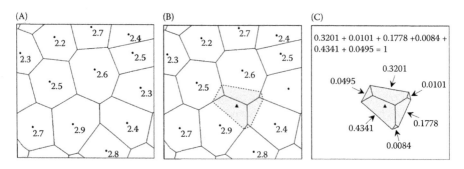

FIGURE 3.6 Natural neighbor interpolation. (A) Voronoi diagram for input points; (B) Output location (triangle symbol); (C) Weights for natural neighbors of the output location.

A new Voronoi polygon is then created at the interpolation point (i.e., output location) (the triangle symbol in Figure 3.6B), and the proportion of overlap between the new Voronoi polygon and the initial Voronoi polygons are used as the weights (Figure 3.6C). A weighted average of neighboring observations (e.g., elevations of LiDAR points) is used as output. The equation for natural neighbor interpolation can be expressed as:

$$g(x, y) = \sum_{i=1}^{n} w_i f(x_i, y_i) \tag{3.2}$$

where $g(x, y)$ is the interpolated value at location (x, y), $f(x_i, y_i)$ are the observations ($i = 1, 2, 3, \ldots, n$), and w_i are the weights. In Figure 3.6, the interpolated value at the triangle is calculated as: $Z_j = 0.3201 \times 2.6 + 0.0101 * 2.3 + 0.1778 \times 2.4 + 0.0084 \times 2.8 + 0.4341 \times 2.9 + 0.0495 \times 2.5 = 2.7$. Additional discussions on natural neighbor interpolation can be found in the works of Li and Heap (2014) and de Smith et al. (2015).

PROJECT 3.1:   CREATING DTM, DSM, AND DHM FROM LiDAR DATA IN INDIANAPOLIS, IN, USA

1. Introduction

   Digital elevation models (DEMs) are the most common digital data representing the surface of the earth. Two common DEM products that can be generated from LiDAR data are DTM and DSM. DTM rasters can be created by spatial interpolation of ground points selected from LiDAR point clouds. DSMs are usually created from the highest LiDAR points representing features on the earth's surface, including buildings, trees, and other objects. Subtracting a DTM from a DSM of the same area can produce a DHM representing the heights of features on the earth's surface, as if these features are placed on a flat, zero-elevation surface. Theoretically, values in a DHM should be zero or positive. However, it is very common to have negative values in DHM due to artifacts, accuracy issues in DTM and DSM, or interpolation processes. Many negative values in DHM are very close to zero, but some others can be substantially less than zero. In this project, methods for creating DTM, DSM, and DHM rasters from LiDAR data and correcting errors in DHM will be introduced. An example of −41 m in DHM caused by laser penetration of glass roofs is also presented. It should be noted that creating DTM, DSM, and DHM is an important first step for many applications, as demonstrated in the projects of the next three chapters.

2. Data

   Public domain LiDAR data (IndianaMap Framework Data: http://www. indianamap.org) for a 1.9 km × 2.1 km area in Indianapolis, IN, USA is used in this project. The LiDAR point clouds were collected in 2011 in a 3-year statewide project with a point density of 1.56 points/m². The horizontal coordinate system is WGS84 [EPSG: 4326], and the vertical coordinate system is North American Vertical Datum 1988 (NAVD88). Ownership of the data products resides with the State of Indiana. This dataset is based on processing services provided by the OpenTopography facility with support from the National Science Foundation under NSF Award Numbers 0930731 and 0930643. The LiDAR data file "Indianapolis.las" can be downloaded (by right-clicking the file and saving it to a local folder) from: http://geography.unt.edu/~pdong/LiDAR/Chapter3/Project3.1/

3. Project Steps

   1. Open an empty Word document so that you can copy any results from the following steps to the document. To copy the whole screen to your Word document, press the PrtSc (print screen) key on your keyboard, then open your Word document and click the "Paste" button or press Ctrl+V to paste the content into your document. To copy an active window to your Word document, press Alt+PrtSc, then paste the content into your document.

   2. Open ArcMap, select Customize → Extensions…, and turn on the Spatial Analyst Extension.

   3. Open ArcToolbox → Data Management Tools → LAS Dataset → Create LAS Dataset. Use Indianapolis.las as input and Indianapolis. lasd as output to create a LiDAR dataset. The LAS dataset is added to ArcMap automatically.

   4. Open the Properties form of Indianapolis.lasd, select the Filter tab, check "Ground" under "Classification Codes", check "All Returns" under "Returns", and then click OK.

   5. Open ArcToolbox → Conversion Tools → To Raster → LAS Dataset to Raster. Select "Indianapolis" from the drop-down list as the input LAS dataset, "dtm" as the output raster in the output folder, ELEVATION as the field value, Binning as the interpolation type, AVERAGE as the cell assignment type, NATURAL_NEIGHBOR as the void fill method, FLOAT as the output data type, CELLSIZE as the sampling type, 1 as the sampling value, and 1 as the Z factor, then click OK to create the output DTM raster (Figure 3.7). Note: You should select the input LAS dataset from the drop-down list because the filter was defined through the layer properties form in Step 4. If you use the browse button to select a LAS dataset as input, all the data points in the LAS files it references will be processed, and the filter defined in Step 4 will not be used.

   6. Open the Properties form of Indianapolis.lasd, select the Filter tab, check "All Classes" under "Classification Codes", check "Return 1" (first return) under "Returns", then click OK. Use "dsm" as the output raster in Step 5 to create the output DSM raster (Figure 3.8).

   7. Open ArcToolbox → Spatial Analyst Tools → Map Algebra → Raster Calculator, use "dsm" − "dtm" as the expression and "dhm" as the output raster in the output folder to create the DHM raster (Figure 3.9). As can be seen in Figure 3.9, the minimum value of the DHM raster is −41.0 m. After identifying the areas with negative values in the DHM raster and comparing the areas with Google Maps images, it is found that most of the negative values are caused by laser penetration of glass roofs.

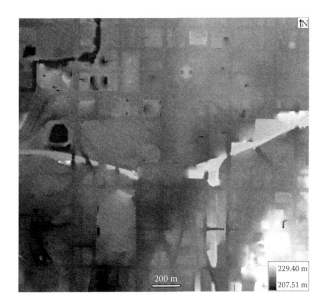

FIGURE 3.7   Output DTM raster.

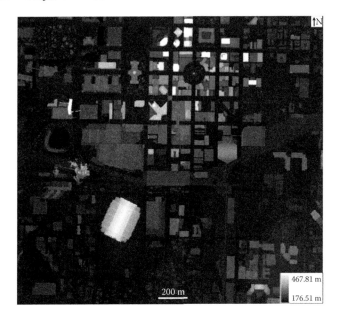

FIGURE 3.8   Output DSM raster.

8. To correct the cell values in the DSM that are lower than the DTM, the CON function in the raster calculator can be used to create a new DSM (dsm2) (Figure 3.10). The syntax of the CON function in Figure 3.10 means: If the DSM value is less than the DTM value, the DTM value will be used in the output new DSM; otherwise the original DSM value

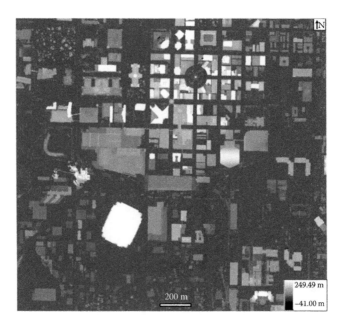

FIGURE 3.9   DHM raster created from DSM and DTM.

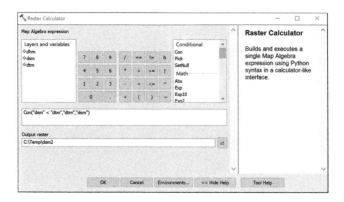

FIGURE 3.10   Calculating a new DSM using the CON function.

is used. Figure 3.11 shows the new DSM, and Figure 3.12 shows the new DHM created by subtracting the DTM from the new DSM.

9. Visualization of the new DHM in ArcScene. Open ArcScene, then click the "Add Data" button to add the new DHM. Open the Layer Properties form of the DHM layer and click the Symbology tab, choose "Stretched", and then select a color ramp and click Apply. Then click the Base Heights tab, choose "Floating on a custom surface," and select the new DHM raster. Then click the button "Raster Resolution…," change Cellsize X and Cellsize Y to 1, and then click OK return to the Layer

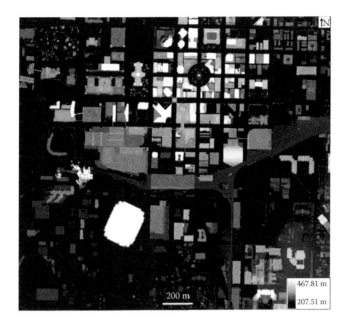

FIGURE 3.11 New DSM obtained from Figure 3.15.

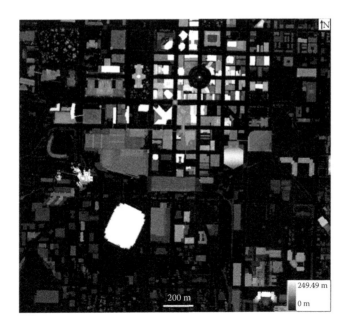

FIGURE 3.12 New DHM obtained from new DSM and DTM.

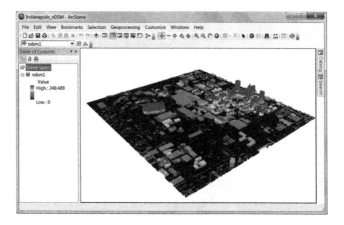

FIGURE 3.13   Visualization of DHM in ArcScene.

Properties form. Click OK to see the results in ArcScene (Figure 3.13). Save the ArcScene project as Indianapolis_DHM.
10. Save your ArcMap and ArcScene projects, and Word document.

PROJECT 3.2:   GENERATING A TERRAIN DATASET USING LiDAR DATA FROM ST. ALBANS, VT, USA

1. Introduction

A TIN is a data model for topographic or non-topographic surfaces based on a connected network of non-overlapping triangles. The nodes of each triangle are the surface points with X, Y, and Z values. A terrain dataset is a multi-resolution TIN stored in a geodatabase. In other words, a terrain dataset has a series of TINs for different map scale ranges. When you are zoomed out to the entire study area, a coarse-resolution TIN is used to display the topographic surface; when you zoom in, increasing levels of detail are available from finer resolution TINs. Therefore, a terrain dataset can facilitate the storage and maintenance of vector-based surface measurements to support spatial modeling in GIS.

In this project, a terrain dataset will be created using LiDAR point clouds from St. Albans in Vermont, USA. First, multipoint feature classes will be created from a LAS file for LiDAR data. Then the feature classes will be added to a feature dataset in a file geodatabase. Finally, new terrain datasets will be created in the feature dataset.

2. Data

LiDAR data from a 2 km × 2 km study area in St. Albans, VT, USA is used in this project. The LiDAR data provided by the U.S. Geological Survey was collected in 2008 with a point density of 2.02 points/m^2. The horizontal coordinate system is Vermont State Plane NAD83 (2007) [EPSG: 32145], and the vertical coordinate system NAVD88 (Geoid 03) [EPSG: 5703]. This dataset is based on processing services provided by the OpenTopography facility with support from the National Science Foundation under NSF Award Numbers 0930731 and 0930643. LiDAR data file in LAS format "St-Albans.las" can be downloaded (by right-clicking the file and saving it to a local folder) from: http://geography.unt.edu/~pdong/LiDAR/Chapter3/Project3.2/

3. Project Steps

1. Open an empty Word document so that you can copy any results from the following steps to the document. To copy the whole screen to your Word document, press the PrtSc (print screen) key on your keyboard, then open your Word document and click the "Paste" button or press Ctrl+V to paste the content into your document. To copy an active window to your Word document, press Alt+PrtSc, then paste the content into your document.

2. Open ArcMap, select Customize → Extensions…, and turn on the 3D Analyst Extension.

3. Create LAS dataset St-Albans.lasd from LAS file St-Albans.las. Open Arctoolbox → Data Management Tools → LAS Dataset → Create LAS Dataset. Use St-Albans.las as input and St-Albans.lasd as output to create a LiDAR dataset in your own folder. The LAS dataset is added to ArcMap automatically. The purpose of this step is to collect coordinate system information that can be used in Step 4 below.

4. Convert the LAS file to multipoint feature class. Go to ArcToolbox → 3D Analyst Tools → Conversion → From File → LAS to Multipoint. Select St-Albans.las as input, dsmpts.shp as output feature class in your project folder, 0.5 as Average Point Spacing, 1 for Input Class Code (Figure 3.14A), and then scroll down and click the icon after the Coordinate System option (Figure 3.14B). On the *XY* Coordinate System panel of the Spatial Reference Properties form (Figure 3.15), select "Import…" to import the *XY* coordinate system from the LAS dataset St-Albans.lasd. For the Z Coordinate System panel in Figure 3.15, choose Vertical Coordinate Systems → North America → NAVD_1988. Then click OK to return to Las to Multipoint conversion form, and click OK. Figure 3.16 shows the attribute table of the multipoint feature class dsmpts.shp.

5. Repeat Step 4, but use dtmpts.shp as output feature class, and 2 for Input Class Code to create a multipoint feature class for DTM.

(A)   (B)

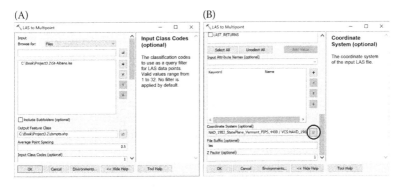

FIGURE 3.14   LAS to multipoint conversion.

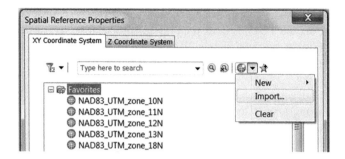

FIGURE 3.15   Selecting coordinate systems for LAS to multipoint conversion.

FID	Shape *	Id	PointCount
0	Multipoint ZM	0	3500
1	Multipoint ZM	0	3500
2	Multipoint ZM	0	3500
3	Multipoint ZM	0	3500
4	Multipoint ZM	0	3500
5	Multipoint ZM	0	3500
6	Multipoint ZM	0	3500
7	Multipoint ZM	0	3500
8	Multipoint ZM	0	3500
9	Multipoint ZM	0	3500
10	Multipoint ZM	0	3500
11	Multipoint ZM	0	3500
12	Multipoint ZM	0	3500

(0 out of 1011 Selected)

dsmpts

FIGURE 3.16   Attribute table of multipoint feature class dmspts.

6. Create file geodatabase. Open ArcCatalog and connect to the folder for this project. Select Customize → Extensions..., and turn on the 3D Analyst Extension for ArcCatalog. Right click the empty space in the project folder in ArcCatalog, and select New → File Geodatabase to create a new file geodatabase "StAlbansGeoDB.gdb".

7. Create new feature dataset "TerrainDS". Right click StAlbansGeoDB. gdb in ArcCatalog and select New → Feature Dataset to create a feature dataset "TerrainDS." Use the coordinate systems explained in Step 4 and default values for other parameters.

8. Import feature classes. Right click TerrainDS and select Import → Feature Class (multiple)... to import dsmpts.shp and dtmpts.shp from the project folder you specified in Step 4, then click OK.

9. Create new Terrain. Right click TerrainDS and select New → Terrain…, then use "StAlbans_Terrain" as the terrain name, and select dsmpts and dtmpts as the feature classes that will participate in the terrain, and 0.5 m as the approximate point spacing, then click Next (Figure 3.17).
10. Use the default values for pyramid type. For the terrain pyramid levels, click "Calculate Pyramid Properties," then click "Next" (Figure 3.18).

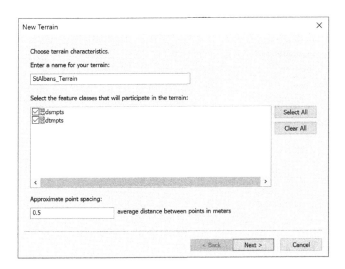

FIGURE 3.17   Creating a new terrain dataset.

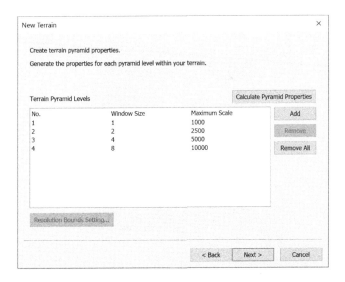

FIGURE 3.18   Creating terrain pyramid properties for a terrain dataset.

11. Click "Finish" on the Summary form to create the terrain dataset "Stalbans_Terrain" (Figure 3.19), then click "Yes" to build the new terrain. Select StAlbans_Terrain in the ArcCatalog Catalog Tree, then click the Preview panel to view the terrain dataset. You can also use Zoom In,

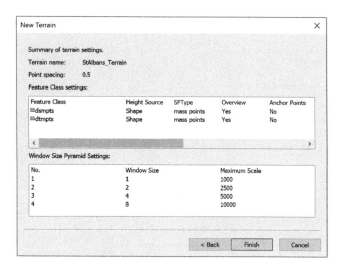

FIGURE 3.19   Summary of terrain settings.

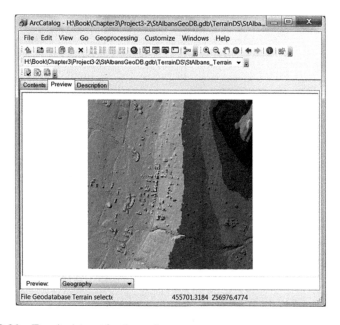

FIGURE 3.20   Terrain dataset for the study area.

Zoom Out, and Pan tools on the ArcCatalog toolbar. Samples of the multi-resolution terrain dataset are shown in Figures 3.20 through 3.22.

12. Repeat Steps 9–11 to create a new Terrain "StAlbansDTM_Terrain" by selecting only the DTM feature class "dtmpts.shp".

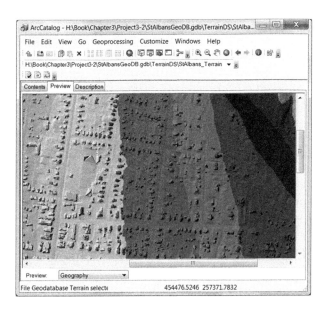

FIGURE 3.21 Details of terrain dataset.

FIGURE 3.22 More details of terrain dataset.

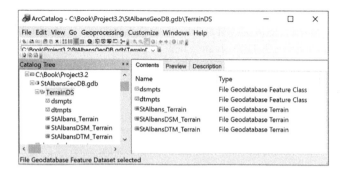

FIGURE 3.23   Five datasets in feature dataset "TerrainDS".

13. Repeat Steps 9–11 to create a new Terrain "StAlbansDSM_Terrain" by selecting only the DSM feature class "dsmpts.shp". At this point, a total of five datasets have been created under the feature dataset "TerrainDS" (Figure 3.23).
14. Save your ArcMap project and Word document.

REFERENCES

Antonarakis, A.S., Richards, K.S., and Brasington, J., 2008. Object-based land cover classification using airborne lidar. *Remote Sensing of Environment*, 112: 2988–2998.

Arefi, H., and Hahn, M., 2005. A morphological reconstruction algorithm for separating off-terrain points from terrain points in laser scanning data. In: *Proceedings of the ISPRS Workshop Laser Scanning*, Enschede, The Netherlands, September 2005.

Auer, S., and Hinz, S., 2007. Automatic extraction of salient geometric entities from LIDAR point clouds. *IEEE International Geoscience and Remote Sensing Symposium (IGARSS 2007)*, 23–28 June 2007, Barcelona, Spain, doi: 10.1109/IGARSS.2007.4423353

Axelsson, P., 1999. Processing of laser scanner data—Algorithms and applications. *ISPRS Journal of Photogrammetry Remote Sensing*, 54: 138–147.

Axelsson, P., 2000. DEM generation from laser scanner data using adaptive TIN models. *The International Archives of the Photogrammetry and Remote Sensing*, 33 (part B4/1): 110–117.

Bork, E.W., and Su, J.G., 2007. Integrating LIDAR data and multispectral imagery for enhanced classification of rangeland vegetation: A meta analysis. *Remote Sensing of Environment*, 111: 11–24.

Chang, Y.-C., Habib, A.F., Lee, D.C., and Yom, J.-H., 2008. Automatic classification of LiDAR data into ground and nonground points. *The International Archives of the Photogrammetry, Remote Sensing and Spatial Information Sciences*, XXXVII (Part B4): 463–468.

Charaniya, A., Manduchi, R., and Lodha, S., 2004. Supervised parametric classification of aerial lidar data. *IEEE Conference on Computer Vision and Pattern Recognition Workshop*, Washington, DC, pp. 30–37.

Chen, Q., 2009. Improvement of the Edge-based Morphological (EM) method for lidar data filtering. *International Journal of Remote Sensing*, 30(4): 1069–1074.

Chen, Q., Gong, P., Baldocchi, D., and Xie, G., 2007. Filtering airborne laser scanning data with morphological methods. *Photogrammetric Engineering and Remote Sensing*, 73: 175–185.

Chen, C., Li, Y., Li, W., and Dai, H., 2013. A multiresolution hierarchical classification algorithm for filtering airborne LiDAR data. *ISPRS Journal of Photogrammetry and Remote Sensing*, 82: 1–9.

Choi, Y.-W., Jang, Y.-W., Lee, H.-J., and Cho, G.-S., 2008. Three-dimensional LiDAR data classifying to extract road point in urban area. *IEEE Geoscience and Remote Sensing Letters*, 5: 725–729.

Cobby, D.M., Mason, D.C., Horritt, M.S., and Bates, P.D., 2003. Two-dimensional hydraulic flood modelling using a finite element mesh decomposed according to vegetation and topographic features derived from airborne scanning laser altimetry. *Hydrological Processes*, 17: 1979–2000.

de Smith, M.J., Goodchild, M.F., and Longley, P.A., 2015. *Geospatial Analysis* (5th edition), Troubador Publishing Ltd, Leicester, UK. http://www.spatialanalysisonline.com/HTML/index.html.

Evans, J.S., and Hudak, A.T., 2007. A multiscale curvature algorithm for classifying discrete return LiDAR in forested environments. *IEEE Transactions on Geoscience and Remote Sensing*, 45: 1029–1038.

Flood, M., 2001. LIDAR activities and research priorities in the commercial sector. *International Archive of Photogrammetry and Remote Sensing*, 34: 3–7.

Haala, N., and Brenner, C., 1999. Extraction of buildings and trees in urban environments. *ISPRS Journal of Photogrammetry and Remote Sensing*, 54: 130–137.

Haugerud, R.A., and Harding, D.J., 2001. Some algorithms for virtual deforestation (VDF) of LIDAR topographic survey data. *IAPRS*, XXXIV (3/W4): 211–218.

Hui, L., Di, L., Xianfeng, H., and Deren, L., 2008. Laser intensity used in classification of lidar point cloud Data. In: *IEEE International Conference on Geoscience and Remote Sensing*, 7–11 July, Boston, pp. 1140–1143.

Kashani, A.G., Olsen, M.J., Parrish, C.E., and Wilson, N., 2015. A review of LiDAR radiometric processing: From ad hoc intensity correction to rigorous radiometric calibration. *Sensors*, 15: 28099–28128.

Kobler, A., Pfeifer, N., Ogrinc, P., Todorovski, L., Ostir, K., and Dzeroski, S., 2007. Repetitive interpolation: A robust algorithm for DTM generation from aerial laser scanner data in forested terrain. *Remote Sensing of Environment*, 108: 9–23.

Kraus, K., and Pfeifer, N., 1998. Determination of terrain models in wooded areas with airborne laser scanner data. *ISPRS Journal of Photogrammetry and Remote Sensing*, 53: 193–203.

Li, J., and Heap, A., 2008. A review of spatial interpolation methods for environmental scientists. *Geoscience Australia*, Record 2008/23, p. 137.

Li, J., and Heap, A.D., 2014. Spatial interpolation methods applied in the environmental sciences: A review. *Environmental Modelling and Software*, 53: 173–189.

Lin, X., and Zhang, J., 2014. Segmentation-based filtering of airborne LiDAR point clouds by progressive densification of terrain segments. *Remote Sensing*, 6: 1294–1326.

Lin, C.-H., Chen, J.-Y., Su, P.-L., and Chen, C.-H., 2014. Eigen-feature analysis of weighted covariance matrices for LiDAR point cloud classification. *ISPRS Journal of Photogrammetry and Remote Sensing*, 94: 70–79.

Liu, X., 2008. Airborne LiDAR for DEM generation: Some critical issues. *Progress in Physical Geography*, 32: 31–49.

Lodha, S., Kreps, E., Helmbold, D., and Fitzpatrick, D., 2006. Aerial LiDAR data classification using support vector machines (SVM). In: *Proceedings of the International Symposium on 3D Data Processing Visualization, and Transmission*, IEEE, Chapel Hill, NC, pp. 567–574.

Mallet, C., Bretar, F., Roux, M., Soergel, U., and Heipke, C., 2011. Relevance assessment of full-waveform LiDAR data for urban area classification. *ISPRS Journal of Photogrammetry and Remote Sensing*, 66: S71–S84.

Meng, X., Currit, N., and Zhao, K., 2010. Ground filtering algorithms for airborne LiDAR data: A review of critical issues. *Remote Sensing*, 2: 833–860.

Mongus, D., and Žalik, B., 2012. Parameter-free ground filtering of LiDAR data for automatic DTM generation. *ISPRS Journal of Photogrammetry and Remote Sensing*, 67: 1–12.

Nie, S., Wang, C., Dong, P., Xi, X., Luo, S., and Qin, H., 2017. A revised progressive TIN densification for filtering airborne LiDAR data. *Measurement*, 104: 70–77.

Niemeyer, J., Rottensteiner, F., and Soergel, U., 2014. Contextual classification of lidar data and building object detection in urban areas. *ISPRS Journal of Photogrammetry and Remote Sensing*, 87: 152–165.

Pingel, T.J., Clarke, K.C., and McBride, W.A., 2013. An improved simple morphological filter for the terrain classification of airborne LIDAR data. *ISPRS Journal of Photogrammetry and Remote Sensing*, 77: 21–30.

Scheiblauer, C., 2014. Interactions with gigantic point clouds. Dissertation submitted in partial fulfillment of the requirements for the degree of Doktor der technischen Wissenschaften, Faculty of Informatics, Vienna University of Technology, 203 p.

Secord, J., and Zakhor, A., 2007. Tree detection in urban regions using aerial lidar and image data. *IEEE Geoscience and Remote Sensing Letters*, 4: 196–200.

Shen, Y., Lindenbergh, R., and Wang, J., 2016. Change analysis in structural laser scanning point clouds: The baseline method. *Sensors*, 17(1): 26, doi: 10.3390/s17010026.

Sibson, R., 1981. A brief description of natural neighbour interpolation. In: (V. Barnett, ed.) *Interpreting Multivariate Data*, Wiley, New York, pp. 21–36.

Sithole, G., and Vosselman, G., 2004. Experimental comparison of filter algorithms for bare-Earth extraction from airborne laser scanning point clouds. *ISPRS Journal of Photogrammetry and Remote Sensing*, 59: 85–101.

Stassopoulou, A., and Caelli, T., 2000. Building detection using Bayesian networks. *International Journal of Pattern Recognition and Artificial Intelligence*, 83: 715–733.

Tobler, W.R., 1970. A computer movie simulating urban growth in the Detroit region. *Economic Geography*, 46: 234–240.

Tóvári, D., and Pfeifer, N., 2005. Segmentation based robust interpolation—A new approach to laser data filtering. *International Archives of Photogrammetry, Remote Sensing and Spatial Information Sciences*, 36: 79–84.

Vosselman, G., 2000. Slope based filtering of laser altimetry data. *International Archives of Photogrammetry and Remote Sensing*, 33: 935–942.

Wang, C., and Tseng, Y., 2010. DEM generation from airborne LiDAR data by adaptive dual-directional slope filter. *International Archives of Photogrammetry and Remote Sensing*, 38: 628–632.

Yan, W.Y., Shaker, A., and El-Ashmawy, N. 2015. Urban land cover classification using airborne LiDAR data: A review. *Remote Sensing of Environment*, 158: 295–310.

Zhang, J., and Lin, X., 2013. Filtering airborne LiDAR data by embedding smoothness constrained segmentation in progressive TIN densification. *ISPRS Journal of Photogrammetry and Remote Sensing*, 81: 44–59.

Zhang, W., Qi, J., Wan, P., Wang, H., Xie, D., Wang, X., and Yan, G., 2016. An easy-to-use airborne LiDAR data filtering method based on cloth simulation. *Remote Sensing*, 8: 501, doi: 10.3390/rs8060501.

4 LiDAR for Forest Applications

4.1 INTRODUCTION

When a laser pulse hits a tree, its energy is reflected back to the sensor or scattered away from the crown surface, and the remaining component is transmitted through foliage gaps. The transmitted energy can be further reflected and scattered by foliage, branches, and stems at lower heights. The position and intensity of the reflected energy peaks along the path of travel are related to the vertical and horizontal structure of the canopy. LiDAR is a revolutionary remote sensing technology for vegetation mapping because (1) laser returns can be used to directly estimate information on vertical structure of vegetation, and (2) conventional optical and radar remote sensing signals are not very sensitive to variations in vertical vegetation structure compared with LiDAR.

Numerous studies have been conducted on small and large footprint LiDAR applications in forest studies, including discrete return LiDAR data (e.g., Chen et al. 2006, Hudak et al. 2006, Jensen et al. 2008, Wang and Glenn 2008, 2009, Evans et al. 2009), full-waveform LiDAR data (e.g., Duncanson et al. 2010, Lefsky 2010, Fieber et al. 2015, Cao et al. 2016, Nie et al. 2016), and fusion of LiDAR and other remotely sensed data (e.g., Hyde et al. 2006, Erdody and Moskal 2010, Jones et al. 2010, Puttonen et al. 2010, Clark et al. 2011, Sun et al. 2011). Some forest characteristics such as canopy height, subcanopy topography, and vertical distribution of intercepted surfaces can be directly estimated from LiDAR data, whereas some other characteristics such as basal area, canopy volume, leaf area index (LAI), and aboveground biomass are usually estimated through modeling. Reviews of LiDAR for forest applications can be found in the work of Lim et al. (2003), Hyyppä et al. (2008), Mallet and Bretar (2009), Wulder et al. (2012), Balenović et al. (2013), Chen (2013), Man et al. (2014), and Lu et al. (2016). Figure 4.1 shows basic categories of LiDAR for forest applications.

The information extraction methods for LiDAR applications in forests can be broadly classified into two categories: individual tree based (Chen et al. 2006; Yu et al. 2010, Duncanson et al. 2014, Latifi et al. 2015) and area-based (Means et al. 2000, Hudak et al. 2008, Yu et al. 2010, Latifi et al. 2015). This chapter introduces several topics on LiDAR applications in forest studies, including (1) canopy surface height modeling and mapping; (2) LiDAR metrics for vegetation modeling; (3) individual tree isolation and mapping; (4) area-based modeling and mapping; and (5) modeling, mapping, and estimating biomass. At the end of the chapter, two step-by-step projects are designed in ArcGIS for extracting canopy heights from leaf-on and leaf-off LiDAR data in Susquehanna Shale Hills, PA, USA, and identifying

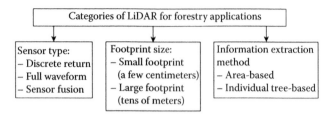

FIGURE 4.1   Categories of LiDAR for forest applications.

disturbances from hurricanes and lightning strikes to mangrove forests using LiDAR data in Everglades National Park, FL, USA.

4.2   CANOPY SURFACE HEIGHT MODELING AND MAPPING

Once a digital terrain model (DTM) is generated using filtered ground returns as described in Chapter 3, useful information about vegetation "height" over forested areas can be immediately derived by subtracting the underlying DTM elevation from the elevation of each point, a process called point cloud normalization or DTM detrending (Figure 4.2). From the detrended LiDAR point cloud, an analyst can create a raster grid of a certain resolution (e.g., 1, 2, 5 m, etc.) and calculate various statistics (e.g., mean, median, and maximum) of height for laser points falling within each grid cell, which results in various vegetation height raster products. In this section, we focus on one particular type of vegetation height product: a raster grid that stores the upper surface (i.e., maximum) height of vegetation canopy, which hereinafter is called the canopy height model (CHM). In our discussion, CHM is essentially 2.5-dimensional (2.5D) instead of 3-dimensional (3D) because, at a given horizontal location, there is only one height; if a user is interested in constructing the complete 3D envelope of tree crowns that includes both the upper surface and lower

FIGURE 4.2   LiDAR point cloud before (left) and after (right) height normalization.

boundary, more sophisticated techniques such as "wrapped surface reconstruction" (Kato et al. 2009) is needed. CHM, as a surface model, by itself carries important information about the amount and spatial distribution of vegetation materials over a geographical area. It is also the basis for mapping individual trees and deriving tree-level information (such as height and crown size) as described in Section 4.4.

A CHM can be generated via two different processes: (1) first generate a digital surface model (DSM) from the original LiDAR point cloud (a point cloud with the raw x, y, z values), and then subtract DTM from DSM to derive CHM, or (2) first create a DTM-detrended point cloud (with x, y, height values) by subtracting each point's Z elevation from its DTM elevation, and then generate a CHM from the detrended point cloud. Whatever method is used, interpolation is needed to generate continuous DSM or CHM models from discrete points.

The main challenge of CHM generation lies in the fact that the LiDAR sensor does not continuously measure but just samples the earth surface (i.e., LiDAR data acquisition is essentially a sampling process). Maximum is an order statistic for which the sample estimate is always less than or equal to the population parameter (i.e., the estimator is biased). Therefore, if CHM is generated by simply searching the laser point of maximum elevation or height within each cell, it would underestimate the true maximum height. This issue has been recognized in numerous studies (e.g., Næsset 1997, Hyyppä and Inkinen 1999, Persson et al. 2002, Holmgren et al. 2003, Gaveau and Hill 2003, Leckie et al. 2003). Another problem with sampling is that some cells, especially when they are small, might not have laser points within them, which appear as "pits" over crowns on a CHM (e.g., Baltsavias 1999, Ben-Arie et al. 2009, Popescu and Wynne 2004). An inaccurate CHM causes problems in the retrieval of individual trees and crown attributes (Nelson et al. 2000, Weinacker et al. 2004, Van Leeuwen et al. 2010), especially when crown shapes are complex and irregular (Mei and Durrieu 2004, Chen et al. 2006, Rahman and Gorte 2009, Ben-Arie et al. 2009, Pitkanen et al. 2004).

Many approaches have been proposed to address the above problems. For example, Van Leeuwen et al. (2010) fitted cone shapes to point clouds to compensate underestimation of tree height. However, such an approach is difficult to apply over forests with trees of different shapes. A better strategy to address these issues is interpolation. Note that choosing an appropriate interpolator is the key. As discussed in Chapter 3, some of the interpolators (e.g., inverse distance weighting – IDW) are exact and cannot make predictions beyond the range of the LiDAR point elevations, which cannot solve the problem of underestimation. Other interpolators such as thin-plate spline (TPS) can generate predictions larger than the maximum elevation of laser points; however, they are sometimes over-sensitive to elevation changes of laser points and thus produce unrealistic predictions.

To reduce or eliminate the effects of data pits, Leckie et al. (2003) proposed a method to assign all LiDAR data into 25 cm × 25 cm grid cell and only the highest points in each cell were retained to generate CHM. However, many pits still remain because the highest points in some grid cells are in fact data pits. Some other studies have focused on post-processing of DSM or CHM to fill the pits using local functions, such as using the median value in a neighborhood to replace a local pixel value (Hyyppä et al. 2000), or filtering pits using mean values (MacMillan et al. 2003).

While these local filtering functions can remove some pits, they also modify the crown shapes. Ben-Arie et al. (2009) proposed a semi-automated pits filling algorithm to identify and fill pits in a CHM and reported that the pit filling algorithm was visually superior to the tested smoothing filters (3×3 median, mean, and Gaussian) for filling data pits while preserving the edges, shape, and structure of the CHM. However, the CHM was derived from a DSM and a DTM which were created using IDW interpolation of the original point clouds, and any subsequent processing of the DSM, DTM, or CHM will not correct the errors induced by the interpolation process. Therefore, accurate representation of canopy surfaces is needed to ensure high quality of a CHM.

Liu and Dong (2014) proposed a new method for generating CHMs from discrete-return LiDAR point clouds to address the common problems of tree height underestimation and data pits in LiDAR-derived CHMs. Unlike the method by Leckie et al. (2003) which uses the highest points in grid cells (here the method is referred to as the highest points method), the method proposed by Liu and Dong takes into account the height distribution of LiDAR points in selection circles. In contrast to the semi-automated algorithm proposed by Ben-Arie et al. (2009) for interpolated raster surfaces, the Liu and Dong method is based on the original LiDAR point clouds in order to minimize the propagation of errors. Figure 4.3 shows the flowchart for pre-processing LiDAR point clouds. Figure 4.4 shows CHMs for simulated single tree LiDAR points generated using the method proposed by Dong (2010) (also see Equations 4.3 and 4.4). Results from real LiDAR data are shown in Figures 4.5 through 4.8. As can be seen from the results, the method proposed by Liu and Dong produced better CHMs in comparison with some other methods.

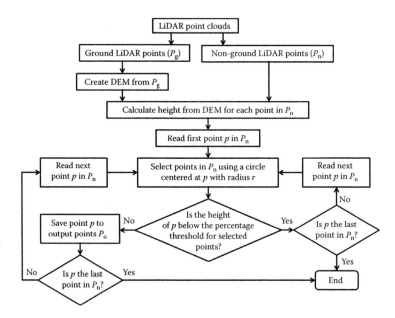

FIGURE 4.3 Flowchart for creating new point clouds P_o. (From Liu, H. and Dong, P., *Remote Sens. Lett.*, 5, 575–582, 2014.)

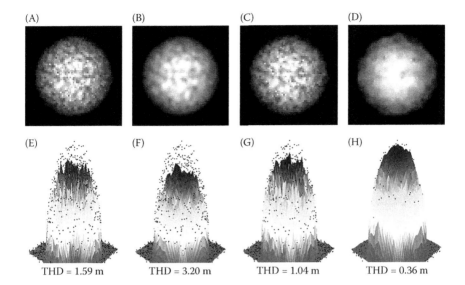

FIGURE 4.4   Grey-scale view (top) and perspective view (bottom) of simulated LiDAR point clouds for simulated tree crowns. (A) IDW interpolation of original non-ground points; (B) Median filtering of (A); (C) IDW interpolation using the highest points method; (D) IDW interpolation using the new method; and (E) through (H) are corresponding perspective views of (A) through (D) along with original non-ground point clouds. THD is the treetop height difference between the treetops in the point clouds and on the interpolated raster surfaces. (From Liu, H. and Dong, P., *Remote Sens. Lett.*, 5, 575–582, 2014.)

As explained earlier, CHM can also be generated as the difference between DSM and DTM. Therefore, accurate construction of DSM can help address the problems of height underestimation and "pit" in raster data. Tang et al. (2013) used a region-based level set method to reconstruct 3D canopy surfaces. Based on the level set function $\phi(t, x, y)$, the zero-level set segments the image into several homogeneous regions by minimizing the following function F (Chan and Vese 2001):

$$
\begin{aligned}
F(c_1, c_2, \varphi) = {}& \mu \int_{\Omega} \delta(\varphi(x,y)) \, |\nabla \varphi(x,y)| \, dx \, dy \\[6pt]
& + v \int_{\Omega} H(\varphi(x,y)) \, dx \, dy \\[6pt]
& + \lambda_1 \int_{\Omega} |u_0(x,y) - c_1|^2 H(\varphi(x,y)) \, dx \, dy \\[6pt]
& + \lambda_2 \int_{\Omega} |u_0(x,y) - c_2|^2 \, (1 - H(\varphi(x,y))) \, dx \, dy
\end{aligned}
\tag{4.1}
$$

where Ω is a bounded open subset of $\mathbb{R}^2$; $u_0(x, y)$ is a given image; (x, y) are pixel coordinates; c_1 and c_2 are the average of u_0 inside and outside the evolving curve, $H(\cdot)$

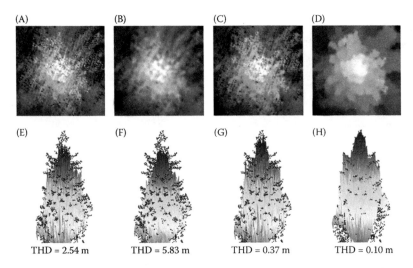

(A) (B) (C) (D)

(E) (F) (G) (H)

THD = 2.54 m   THD = 5.83 m   THD = 0.37 m   THD = 0.10 m

FIGURE 4.5 Grey-scale view (top) and perspective view (bottom) of LiDAR point clouds for a Douglas fir crown. (A) IDW interpolation of original non-ground points; (B) Median filtering of (A); (C) IDW interpolation using the highest points method; (D) IDW interpolation using the new method. (E) through (H) are corresponding perspective views of (A) through (D) along with original point clouds. THD is the treetop height difference between the treetops in the point clouds and on the interpolated raster surfaces. (From Liu, H. and Dong, P., *Remote Sens. Lett.*, 5, 575–582, 2014.)

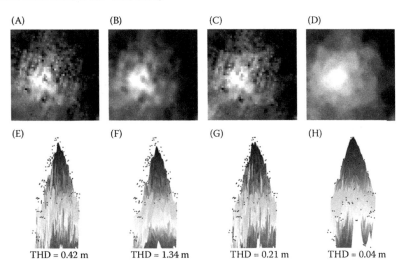

(A) (B) (C) (D)

(E) (F) (G) (H)

THD = 0.42 m   THD = 1.34 m   THD = 0.21 m   THD = 0.04 m

FIGURE 4.6 Grey-scale view (top) and perspective view (bottom) of LiDAR point clouds for a Douglas fir crown. (A) IDW interpolation of original non-ground points; (B) Median filtering of (A); (C) IDW interpolation using the highest points method; (D) IDW interpolation using the new method. (E) through (H) are corresponding perspective views of (A) through (D) along with original point clouds. THD is the treetop height difference between the treetops in the point clouds and on the interpolated raster surfaces. (From Liu, H. and Dong, P., *Remote Sens. Lett.*, 5, 575–582, 2014.)

is the Heaviside function; and $\delta(\cdot)$ is the Kronecker delta function. The first element in Equation 4.1 is the length of the contour, the second is the area within the contour, and the third and fourth terms are proportional to the energy inside and outside the contour.

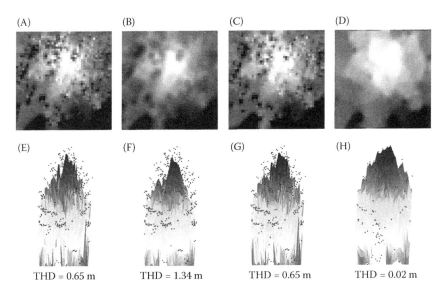

FIGURE 4.7 Grey-scale view (top) and perspective view (bottom) of LiDAR point clouds for a redwood crown. (A) IDW interpolation of original non-ground points; (B) Median filtering of (A); (C) IDW interpolation using the highest points method; (D) IDW interpolation using the new method. (E) through (H) are corresponding perspective views of (A) through (D) along with original point clouds. THD is the treetop height difference between the treetops in the point clouds and on the interpolated raster surfaces. (From Liu, H. and Dong, P., *Remote Sens. Lett.*, 5, 575–582, 2014.)

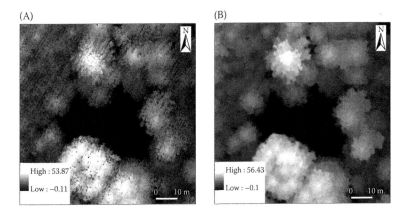

FIGURE 4.8 Comparison of canopy height models created from IDW interpolation of (A) all points and (B) remained points filtered by the new method. (From Liu, H. and Dong, P., *Remote Sens. Lett.*, 5, 575–582, 2014.)

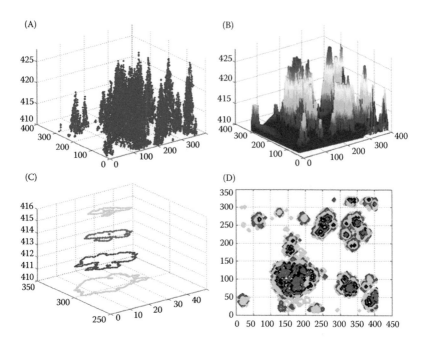

FIGURE 4.9   Region-based level set for 3D tree canopy reconstruction. (A) Perspective view of raw LiDAR data; (B) Constructed canopy surface; (C) Stack of slices for a group of trees; and (D) 2D section of canopy at different heights. (From Tang, S. et al., *Int. J. Remote Sens.*, 34, 1373–1385, 2013.)

The minimization of function F in Equation (4.2) is solved by taking the Euler-Lagrange equations and updating the level set function ϕ by gradient descent:

$$\frac{\partial \phi}{\partial t} = \delta(\phi)\left[\mu \cdot div\left(\frac{\nabla \phi}{|\nabla \phi|}\right) - \lambda_1(\mu_0 - c_1)^2 + \lambda_2(\mu_0 - c_2)^2\right] \qquad (4.2)$$

Figure 4.9 shows some results of 3D tree canopy construction using the region-based level set method. The reconstructed 3D canopy surfaces allow for more accurate measurement of tree crown parameters, such as crown volume and radius.

4.3   LiDAR METRICS FOR VEGETATION ANALYSIS

CHM described in the previous section is to characterize the canopy surface "maximum" height over space. However, over any particular area (e.g., a 1 m × 1 m grid cell, a tree crown, or a circular field plot), many laser points usually exist. Based on the height of these points, several statistical metrics (e.g., mean and standard deviation) can be generated. These metrics can be used to characterize the 3D canopy structure and predict forest attributes that are needed for many practical applications (e.g., forest management, ecological modeling, and climate policy analysis).

It is important to note that LiDAR does not directly "measure" most of the vegetation attributes that are needed in practice, such as basal area, canopy cover, stand density, and biomass. Even for vegetation height, the LiDAR measurements are point-based while field-based measurements are tree-based. In other words, LiDAR- and field-based height measurements are at different scales. To address these discrepancies, a common approach is to develop statistical (e.g., regression) models to indirectly "model" or predict the forest attributes over a geographical area using LiDAR-based metrics.

Based on previous studies (Means et al. 2000, Hudak et al. 2006, 2008), Evans et al. (2009) proposed 22 metrics for vegetation modeling. The metrics are listed below, where x = numeric variable, n = number of observations, μ = mean, and σ = standard deviation. Note that, however, this is not a standard list of LiDAR metrics for vegetation analysis. Numerous alternative ways exist to calculate LiDAR metrics; for example, percentile heights can be calculated at any value (e.g., 90th, 80th, and 70th). Whether a LiDAR metric is optimal for predicting a forest attribute depends on many factors such as the type of statistical models used (e.g., parametric vs. non-parametric models), the scale of analysis (e.g., individual trees or plots), and the nature of the attribute (e.g., whether it is 1D canopy height or 3D volume). A user should consider the forest attribute of interest and carefully choose an appropriate way to calculate LiDAR metrics.

1. Minimum;
2. Maximum;
3. Range: (max − min);
4. Arithmetic Mean: $\mu = \dfrac{\sum_{i=1}^{N} x_i}{N}$
5. Standard Deviation (σ);
6. Variance (σ^2);
7. Percentiles: 5th, 10th, 25th, 50th, 75th, and 95th percentile values (x);
8. Median Absolute Deviation from Median: $median_i(X_i - median_j(X_j))$;
9. Dominate Mode: Value of the dominate mode in a kernel density estimate (x);
10. Skewness: $\dfrac{\sqrt{n} \sum_{i=1}^{n} (x_i - \mu)^3}{\left(\sum_{i=1}^{n} (x_i - \mu)^2 \right)^{3/2}}$
11. Kurtosis: $\dfrac{\frac{1}{n} \sum_{i=1}^{n} (x_i - \mu)^4}{\left(\frac{1}{n} \sum_{i=1}^{n} (x_i - \mu)^2 \right)^2} - 3$
12. Interquartile range: 75th percentile (x) − 25th percentile (x);
13. Coefficient of Variation: $(\sigma/\mu) \cdot 100$;
14. Number of Modes: Number of modes from a kernel density estimate (x);
15. Difference between Min and Max Mode: (maximum mode − minimum mode) from a kernel density estimate (x);

16. Canopy Relief Ratio: $\dfrac{\mu\,(height) - \text{Min}\,(height)}{\text{M}\,\text{ax}\,(height) - \text{Min}\,(height)}$

17. Percent of returns that are first, second, third, etc.: $\dfrac{\left[\,n_i\,\middle|\,return\ number\,\right]}{N} \cdot 100$

18. Texture: $\sigma(n_i\ |> height(0)\ \text{and} <= height(1)|)$;

19. Number of ground returns;

20. Number of non-ground returns;

21. Density: $\dfrac{\left[\,n_i\,\middle|\,nonground\,\middle|\,\right]}{N} \cdot 100$

22. Stratified Density: $\dfrac{\left[\,n_i\,\middle|\,> x_1\ and < x_2\,\middle|\,\right]}{N} \cdot 100$

4.4 INDIVIDUAL TREE ISOLATION AND MAPPING

In forestry and ecology studies, tree-level information (e.g., height, crown size, diameter at breast height (DBH), stem volume, and biomass) is often needed. For decades, researchers have attempted to map individual trees using remote sensing data; initially with aerial photos (e.g., Leckie et al. 1992; Wulder et al. 2000) and later with high spatial resolution satellite imagery (e.g., Chopping 2011). Airborne LiDAR data of high point density (~5 points/m² or higher) can produce better results than high spatial resolution optical imagery because (1) all points in a LiDAR point cloud are accurately geo-referenced in 3D whereas pixels in an image or photo are projected to 2D with distortion (such as relief displacement), (2) the basis of detecting trees from LiDAR data is the 3D shape of the trees whereas photo- or image-based tree detection replies on the pixel brightness variations within tree crowns (the latter is often affected by sun illumination conditions), and (3) LiDAR can provide direct estimates of tree height whereas it is much more difficult to do so using imagery. Because of these advantages, airborne LiDAR has gained popularity for tree mapping in the 21st century.

Numerous methods of mapping trees from LiDAR data have been developed, and they can be broadly classified into categories based on either (1) 2D or 3D grid models of canopy (e.g., CHM) or (2) 3D point clouds. The developed methods differ in the complexity of algorithms, ease of setting up or tuning parameters, computation resource needed, processing speed, and type of trees (dominant trees only versus all trees including understory ones) that can be extracted. The rest of this section will introduce the grid- and point-based tree isolation algorithms, respectively, and discuss their strengths and weaknesses.

4.4.1 GRID-BASED TREE MAPPING

The classical approaches to map individual trees are based on CHM, a 2D grid or raster of canopy height. A common strategy is to identify treetops by searching local maxima from CHM (Persson et al. 2002, Leckie et al. 2003, Maltamo et al. 2004, Popescu and Wynne 2004, Monnet et al. 2010, Vastaranta et al. 2011). Several challenges exist to detect trees from local maxima approaches: (1) trees of irregular shape could have hanging branches that appear as local maxima in the CHM; this is quite common for old, large deciduous trees, especially at leaf-off stage; and (2) the "pits" mentioned in Section 4.2 could create artificial low valleys and thus local maxima in CHM.

The local maxima that correspond to false treetops can be suppressed using two strategies: (1) smoothing CHM using low-pass filter to remove the "pits", and (2) using variable window sizes for searching local maxima. Note that the minimal window size of searching local maxima is 3×3 pixels. The latter strategy means that the window size can be increased according to tree size. For example, Popescu and Wynne (2004) adaptively varied the window size with the height of each CHM pixel, according to a regression curve that can characterize the tree height-crown size relationship at a given site (Figure 4.10). The regression curve was calibrated from a field sample of trees for which tree height and crown radius were measured. Since larger window sizes were used to search local maxima for taller canopy, fewer false treetops were included in the results. In other words, the commission errors in treetop detection were reduced using variable window size (see the example in the top row of Figure 4.11).

Chen et al. (2006) reported that although the use of variable window size based on regression curve can reduce the commission errors, it can also introduce omission errors in treetop detection. This is because the window size based on the regression curve is larger than crown size for about half of the trees (see Figure 4.10). When the window size is larger than the crown size, some true treetops might be missed (see the bottom row of Figure 4.11).

To detect the missed treetops, Chen et al. (2016) used the lower-limit of the prediction interval of the regression model (Figure 4.10). This indeed can reduce the

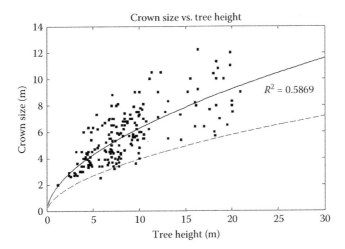

FIGURE 4.10 Variable window size for searching local maxima based on either the regression curve (the solid line) or the lower limit of the prediction interval of the regression model (the dashed line). (Adapted from Chen, Q. et al., *Photogramm. Eng. Remote Sens.*, 72, 923–932, 2006.)

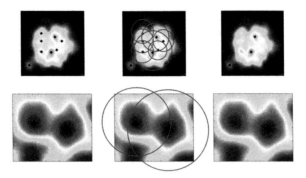

FIGURE 4.11 Local maxima (dots) detected using fixed (left) and variable window size (right). The circles in the middle figures shows the variable window sizes used at each local maximum detected from fixed window size. The top figures are for a case of two field trees (one small in the lower-left corner, one very large tree in the middle); the bottom figures are for two trees that are tall but with small crown sizes.

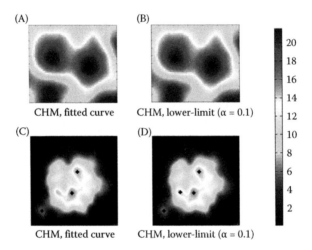

FIGURE 4.12 Treetops (dots) detected when window sizes vary with regression curves (A and C) or lower-limit of the prediction intervals of the regression model (B and D). α is the significance level.

omission errors in treetop detection for some trees (see Figure 4.12B), but it will slightly increase the commission errors for others (Figure 4.12D). To reduce commission and omission errors simultaneously, Chen et al. (2006) first smoothed the CHM with a maximum filter with variable window size (similar to the lower limit of the prediction interval in Figure 4.10 but with a different significance level). The resulted model was named Canopy Maximum Model (Figure 4.13), which was used to detect treetops with variable window size based on the lower-limit curve.

Using such a divide-and-conquer strategy, Chen et al. (2006) found that treetops can be properly detected (Figure 4.14A) while using variable window size based on

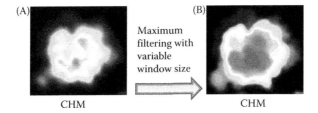

(A)   (B)

Maximum
filtering with
variable
window size

CHM   CHM

FIGURE 4.13 Generation of canopy maximum model (B) from CHM (A) by applying maximum filtering with variable window size.

CHM (Figure 4.14B and D), but cannot be properly detected when using fixed window size (Figure 4.14C).

Since treetop detection can provide (x, y) coordinates of individual trees, tree heights can be also estimated using the CHM value at the location of treetops. However, additional processing is needed to extract crown size, another key attribute of individual trees. Some methods can only extract crown size along limited directions. For example, Popescu et al. (2003) estimated crown radii by fitting four-order polynomial models to two perpendicular directions from treetops. Other methods used image segmentation to create crown segments for individual trees (Hyyppä et al. 2002, Morsdorf et al. 2003) and return the boundary of each crown. Among the various image segmentation methods, the marker-controlled watershed segmentation is probably the most popular one (Koch et al. 2003, Chen et al. 2006, Kwak et al.

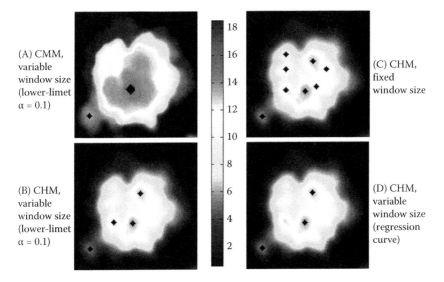

(A) CMM, variable window size (lower-limet $\alpha = 0.1$)

(B) CHM, variable window size (lower-limet $\alpha = 0.1$)

(C) CHM, fixed window size

(D) CHM, variable window size (regression curve)

18
16
14
12
10
8
6
4
2

FIGURE 4.14 Searching local maxima for detecting treetops using (A) CMM and variable window size based on lower limit curve; (B) CHM and variable window size based on lower-limit curve; (C) CHM and fixed window size; and (D) CHM and variable window size based on regression curve. (Adapted from Chen, Q. et al., *Photogramm. Eng. Remote Sens.*, 72, 923–932, 2006.)

2007, Breidenbach et al. 2010, Hu et al. 2014). Using CHM as an example, this algorithm can be understood intuitively as follows: punch a CHM with holes at the location of treetops (which is called markers), invert the punched CHM and immerse it into water, and then build dams at the location where water from different crowns (watersheds) meet (Figure 4.15); the dams are used to delineate the boundary of tree crowns.

The key for the success of marker-controlled watershed segmentation is that treetops (markers) should be properly detected. Commission and omission errors in treetop detection will lead to over- and under-segmentation of tree crowns, respectively. Chen et al. (2006) found that a laser scanner may miss the treetops so local maxima do not exist on a CHM for some trees. Based on the fact that treetops usually are horizontally near the center of tree crowns, Chen et al. (2006) used a distance-transformed image to detect the missing treetops (Figure 4.16).

With a CHM, it is relatively easy to detect large overstory dominant/co-dominant trees, but it is nearly impossible to find smaller understory suppressed trees. To address this problem, many different approaches have been proposed in recent years. For example, Duncanson et al. (2014) proposed a multi-layered crown delineation algorithm as follows: first create tree crown segments using watershed segmentation based on CHM, then analyze the point cloud within each segment, find the point cloud for understory trees, generate a CHM using the understory point cloud, and finally apply marker-controlled watershed segmentation to the understory CHM for detecting trees.

Besides watershed segmentation, other segmentation methods were also used for tree crown mapping. For example, Reitberger et al. (2009) created 3D grid of voxels for canopy and used the normalized cut segmentation to detect both overstory and understory trees; Wang et al. (2008) used a voxel structure and a hierarchical morphological algorithm to generate a crown region at different height intervals. Tree crowns are then reconstructed using a pre-order forest traversal approach, allowing the grouping of neighboring crown regions generated at the same height intervals. The method identifies both canopy and over-topped trees but is sensitive to both the voxel scale and the size of morphological elements (Wang et al. 2008). Liu et al. (2015) proposed a new approach for individual tree crown delineation through crown boundary refinement based on a Fishing Net Dragging method and segment merging based on boundary classification; Strîmbu and Strîmbu (2015) presented a graph-based segmentation algorithm for extracting tree crowns from LiDAR data.

FIGURE 4.15 Marker-controlled watershed segmentation. (Left) CHM, (middle) inverted CHM, (right) inverted CHM with dam built to separate the watersheds. Dark markers are treetops. (Adapted from Chen, Q. et al., *Photogramm. Eng. Remote Sens.*, 72, 923–932, 2006.)

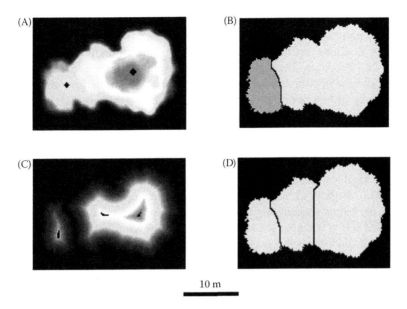

10 m

FIGURE 4.16 Treetop detection and crown delineation using distance-transformed image. (A) CHM for which local maximum does not exist for the middle tree; (B) marker-controlled watershed segmentation using CHM and the two treetops based on local maxima; (C) distance-transformed image based on the results in (B) and the local maxima (treetops), and (D) final marker-controlled watershed segmentation based on the results in (C). (Adapted from Chen, Q. et al., *Photogramm. Eng. Remote Sens.*, 72, 923–932, 2006.)

4.4.2 POINT-BASED TREE MAPPING

Another strategy of tree mapping is to group laser points into clusters that correspond to individual tree crowns. Such clustering algorithms usually assign a 3D point to its tree crown based on its proximity to the center of the tree crown. Therefore, correctly identifying the centers of tree crowns is the key for the success of such approaches, which is similar to the importance of searching treetops for grid-based methods.

Several clustering approaches have been developed to identify tree crown centers and separate point clouds from individual tree crowns. One such approach is the mean-shift mode-seeking algorithm. All laser points from a tree crown can be assumed to be sample data from a 3D density function, which has its mode or local maximum near the center of the tree crown, both horizontally and vertically. Mean-shift estimates or reconstructs the density function using a smoothing function (also called kernel function), for which the key parameter is the kernel bandwidth. A bandwidth that is too large or too small will create a density function that is too smooth or too rough and thus can lead to under- or over-segmentation of point cloud for tree mapping. In this sense, the bandwidth is similar to the window size for smoothing CHM using low-pass filters for grid-based approaches. Ferraz et al. (2012) proposed some bandwidth values for a Mediterranean forest in Portugal. However, the wider applicability of these values into other forests remains unclear.

Another approach of detecting tree crown centers and their associated points is to sort and process all points from the highest to the lowest based on their canopy heights or Z values and then sequentially assign them to individual tree crowns. Li et al. (2012) selected the highest point as the initial point of the first tree crown and checked the next highest point, which was assigned to the first tree crown if it was nearby or to a new tree crown if it was far away; this process was repeated until all points were classified. One of the challenges of applying such an approach is to set an appropriate threshold for determining whether a point is close to or far from the existing trees. Li et al. (2012) mentioned that the threshold should be adaptively changed according to tree height, similar to the variable window size used for searching local maxima in grid-based approaches (Popescu and Wynne 2004; Chen et al. 2006). Vega et al. (2014) calculated the k-nearest neighbor for each point, checked the points sequentially from the highest to the lowest Z, and initialized a new cluster if it was higher than its neighbors. However, a k value that is too large or too small can lead to under- or over-segmentation of tree point cloud. Therefore, they proposed to segment trees at several scales with different k values and assess the likelihood that a segment is a tree based on several criteria such as size, circularity, orientation, and regularity. Overall, these methods have the same challenge as the grid-based approaches in that parameters need to be carefully chosen to avoid tree segmentation errors.

Trees can also be detected based on the density of points (Rahman and Gorte 2009; Mongus and Žalik 2015). For example, Mongus and Žalik (2015) used locations of high point density to detect tree trunks, based on which trees were further segmented using watershed segmentation.

4.4.3 REMAINING CHALLENGES IN TREE MAPPING

Despite many years of research, various tree isolation methods are still subject to several major challenges: (1) although it is relatively easy to detect trees in stands of simple canopy structure (e.g., plantations), the accuracy is often low for complex forest conditions (e.g., multi-stemmed trees, old deciduous trees, multi-layered canopy). It is much more difficult to detect smaller suppressed trees than larger dominant/co-dominant trees. (2) The generality of most tree isolation methods needs to be substantially improved. Although many studies reported superior, sometimes surprisingly high, accuracy for their own study sites, the application of these methods at a different site is often less satisfactory. For example, Kaartinen et al. (2012) did an international benchmarking comparison of nine different methods for the same study site in Europe. They found that the methods did not perform much better and often worse than simple local maxima and watershed segmentation-based methods. (3) Few methods have proved their capacity in mapping trees over large areas (e.g., hundreds of kilometers or greater). This is especially the case for the 3D point- or voxel-based approaches, which are very computation-demanding. In contrast, CHM-based approaches are relatively fast; however, they need to improve their performance in detecting overtopped trees. Overall, the wider use of LiDAR to map individual trees for practical forest management relies on further improvements in the accuracy, generality, and computational speed/efficiency in tree isolation algorithms.

4.4.4 Modeling and Predicting Individual Tree Attributes

From the detected individual treetops and delineated tree crowns, an analyst can directly estimate a few key attributes that relate to the position and shape of individual tree crowns, including tree center/top location (x, y, and z), tree height, crown size (radius or area), height to the live crown, and crown depth. However, due to occlusion of LiDAR energy by upper canopy, and even with very high point density, airborne LiDAR data is not dense enough to see the details of trunks, branches, and leaves of trees. LiDAR cannot directly measure properties (such as mass and weight) other than the location and geometry of trees. Therefore, if airborne LiDAR data is used other tree-level attributes such as DBH, basal area, stem volume, leaf area, and biomass have to be indirectly estimated, usually using statistical methods. This requires two types of data: (1) field data for these attributes need to be collected from a sample of trees, and (2) LiDAR metrics, as described in Section 4.3, need to be extracted for each delineated tree crown. The field data and LiDAR metrics for the sampled trees are used to calibrate the models, which are further used to predict the attributes for the remaining trees (not sampled in the field) based on their LiDAR metrics at the crown-level.

The above procedure works only if individual tree crowns are accurately delineated. However, as discussed in the previous section, tree isolation of most forests is prone to errors. If the tree crown boundary is mistakenly delineated, the corresponding tree-level LiDAR metrics will also be incorrect. In such a case, it is unrealistic to expect high accuracy of estimating attributes at the tree level. Before an ideal algorithm of tree isolation is developed, a more plausible direction of research is to determine how the total of the predicted attributes from LiDAR-based segments can be as close as possible to the total of the true attributes of the corresponding trees in the field over an area. This means that the LiDAR metrics and/or statistical models should be carefully chosen to ensure that the estimate of the totals is insensitive to tree segmentation errors. Chen et al. (2007) proposed the use of the geometric volume of tree crown segments (the volume under CHM) to predict basal area and stem volume within forest plots; they also developed theoretical parametric models based on general allometry theory to predict these two attributes using geometric volume. It was found that the theoretical models, even with only one predictor, had better performance than other data-driven empirical models, some of which were developed from over 40 LiDAR metrics. The encouraging results from such a study imply that, to improve prediction accuracy, biology and ecology theories should be utilized to the maximum extent to develop statistical models that are both general and interpretable.

4.4.5 A Simulation Study for Tree Isolation

Since accurate positions of individual trees in dense forests are difficult to obtain using GPS devices, LiDAR point cloud simulation based on known treetop locations can be helpful for development and validation of treetop detection and crown shape analysis methods. Dong (2009) used random points with random elevation variations near the surfaces of three simple geometric models (cone, hemisphere, and half-ellipsoid) to simulate individual tree crowns. Based on observation of discrete return LiDAR point clouds for individual conifer and deciduous tree crowns in the Soquel State Demonstration Forest near Santa Cruz, CA, USA, Dong (2010) noticed that that a majority of LiDAR points distribute in a layer near the crown surface because laser pulses have a reduced probability of hitting the core of the crown, and proposed a revised model for simulating LiDAR point clouds for individual trees (Figure 4.17). Figure 4.17A shows LiDAR points of a conifer tree crown, with red points representing selected LiDAR points from the x-z plane to show the point distribution in a profile. In Figure 4.17B, the outer surface $f_2(x, y)$ of the points is a half-ellipsoid with semi-principal axes of length a, b, c ($a = b = r < c$), whereas the inner surface $f_1(x, y)$ is a half-ellipsoid with $a = b = d$. Theoretically, d can change in the range of $0 < d < r$, depending on the tree species and season (leaf-on or leaf-off). For simplicity, $d = r/2$ is used to construct the model for generating simulation data (Figure 4.17B). Similar diagrams can be drawn for cone and hemisphere models.

Based on the model in Figure 4.17, random points can be generated in a circle with radius r in the x-y plane, and z values of the points can be calculated using the following equations:

$$z(x, y) = f_2(x, y) \cdot t \quad \left(\left(\frac{r}{2} \right)^2 \le x^2 + y^2 \le r^2 \right) \tag{4.3}$$

$$z(x, y) = f_1(x, y) + \left(f_2(x, y) - f_1(x, y) \right) \cdot t \quad \left(x^2 + y^2 < \left(\frac{r}{2} \right)^2 \right) \tag{4.4}$$

where t is a random number between 0 and 1. Equation (4.3) is for points above the shaded ring in Figure 4.17C, and Equation (4.4) is for points above $f_1(x, y)$ and below

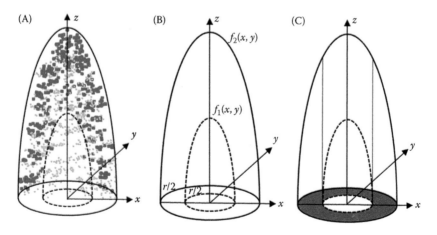

FIGURE 4.17 LiDAR points selected from a crown profile (A), 3D surfaces $f_1(x, y)$ and $f_2(x, y)$ (B), and shaded and non-shaded areas for point simulation (C). (From Dong, P., *Remote Sens. Lett.*, 1, 159–167, 2010.)

$f_2(x, y)$. The surfaces $f_1(x, y)$ and $f_2(x, y)$ can be defined by geometric models such as cone, hemisphere, and half-ellipsoid (Dong 2009). Figure 4.18 shows 12 simulated tree crowns with different radius r and number of points N, and a point density of about 8.8 points/m^2.

Using the three simulated tree crowns of different shapes (cone, hemisphere, and half-ellipsoid) in Figure 4.18D, Dong (2010) tested automated discrimination of different tree crowns using 3D shape signatures (Osada et al. 2002, Dong 2009, 2010). The 3D shape signatures of crowns ($r = 5$ m) obtained from sample circles with different sizes (radius s) and locations are shown in Figure 4.19. Correlation coefficients between the 3D shape signatures are listed in Tables 4.1 through 4.3. The results show that 3D shape signatures calculated from different sizes of sample circles may show different degrees of separability between crown shapes, and that slight deviation (less than 20% of crown width) of the center of sample circles from the crown center does not change the 3D shape signatures notably. The results suggest that it is possible to separate different crown shapes in an automated manner once the treetops are accurately detected.

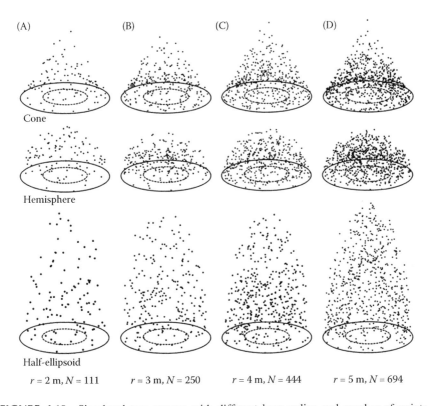

FIGURE 4.18 Simulated tree crowns with different base radius and number of points. (A) $r = 2$ m, $N = 111$; (B) $r = 3$ m, $N = 250$; (C) $r = 4$ m, $N = 444$; (D) $r = 5$ m, $N = 694$; r—radius, N—number of points. (From Dong, P., *Remote Sens. Lett.*, 1, 159–167, 2010.)

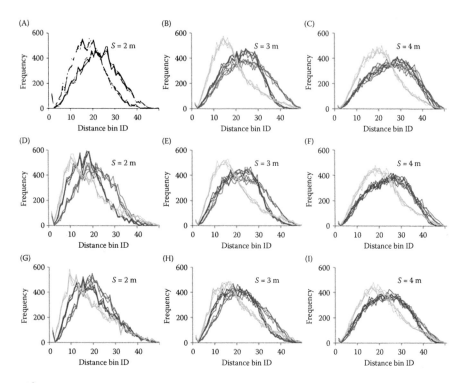

FIGURE 4.19 3D shape signatures of crowns ($r = 5$ m) obtained from sample circles with different sizes (radius s) and locations. First row: results from sample circles around the crown center; second row: results from sample circles centered at a point 1 m away from the crown center; third row: results from sample circles centered at a point 2 m away from the crown center. (Blue—cone, red—hemisphere, green—half-ellipsoid). (From Dong, P., *Remote Sens. Lett.*, 1, 159–167, 2010.)

TABLE 4.1

Correlation Coefficients between 3D Shape Signatures Obtained from Sample Circles (with radius s) Centered at the Crown Center

	$s = 2$ m			$s = 3$ m			$s = 4$ m		
	Cone	Sphr	Elps	Cone	Sphr	Elps	Cone	Sphr	Elps
Cone	1	0.79	0.96	1	0.93	0.67	1	0.97	0.69
Sphr	0.79	1	0.77	0.93	1	0.57	0.97	1	0.78
Elps	0.96	0.77	1	0.67	0.57	1	0.69	0.78	1

Source: Dong, P., *Remote Sens. Lett.*, 1, 159–167, 2010.
Sphr, Semisphere; Elps, Half-ellipsoid.

TABLE 4.2

Correlation Coefficients between 3D Shape Signatures Obtained from Sample Circles (with radius s) Centered at a Point 1 m away from the Crown Center

	s = 2 m			s = 3 m			s = 4 m		
	Cone	Sphr	Elps	Cone	Sphr	Elps	Cone	Sphr	Elps
Cone	1	0.77	0.90	1	0.93	0.81	1	0.97	0.82
Sphr	0.77	1	0.59	0.93	1	0.71	0.97	1	0.86
Elps	0.90	0.59	1	0.81	0.71	1	0.82	0.86	1

Source: Dong, P., *Remote Sens. Lett.*, 1, 159–167, 2010.
Sphr, Semisphere; Elps, Half-ellipsoid.

TABLE 4.3

Correlation Coefficients between 3D Shape Signatures Obtained from Sample Circles (with radius s) Centered at a Point 2 m away from the Crown Center

	s = 2 m			s = 3 m			s = 4 m		
	Cone	Sphr	Elps	Cone	Sphr	Elps	Cone	Sphr	Elps
Cone	1	0.91	0.90	1	0.90	0.90	1	0.98	0.89
Sphr	0.91	1	0.68	0.90	1	0.73	0.98	1	0.84
Elps	0.90	0.68	1	0.90	0.73	1	0.89	0.84	1

Source: Dong, P., *Remote Sens. Lett.*, 1, 159–167, 2010.
Sphr, Semisphere; Elps, Half-ellipsoid.

4.5 AREA-BASED MODELING AND MAPPING

As described in the previous section, isolating and mapping individual trees is technically challenging and prone to errors. It appears few studies have applied LiDAR to map individual trees at large spatial scales. Over areas of tens of square kilometers or larger, area-based approaches are more commonly used. This usually involves the following steps: (1) a sample of forest plots are set up in the field, where forest attributes are measured at the tree level and summarized at the plot level; (2) LiDAR metrics are extracted within these field plots; (3) LiDAR metrics are extracted for the whole study area by partitioning the study area into grid cells of equivalent size to field plots and calculating LiDAR metrics within each grid cell; (4) statistical models are developed to predict forest attributes using LiDAR metrics using the data at the plot level; and (5) the developed plot-level models are applied to each grid cell to predict and map forest attributes for the whole study area. Compared to tree-based approaches, more field work is involved because field data are collected at the plot level, not at the tree level;

however, the relevant LiDAR data processing is greatly simplified by just calculating LiDAR metrics within individual plots and raster grid cells.

Caution should be exercised for several issues regarding area-based approaches. First, since the relationship is nonlinear and scale-dependent between LiDAR metrics and most forest attributes, the grid cell size for model prediction (i.e., minimal mapping unit) has to be equivalent to the field plot size. A critical issue is to choose the proper field plot size. When calculating plot-level forest attributes using tree-level attributes, the common practice is that a tree is included or excluded based on whether the tree trunk is inside or outside of the plot, even if the tree crown is partially inside the plot. However, the extraction of LiDAR metrics typically uses cookie-cutter approaches, so inconsistency exists between LiDAR metrics and field data for trees near the edge of plots. Such "edge-effects" are more severe for small plots, leading to large errors in modeling forest attributes (Frazer et al. 2011). One way of reducing modeling errors is to use larger plots, which, however, also requires more field work per plot. The choice of proper field plot size and, in general, field plot design, is an important yet difficult issue.

Second, LiDAR metrics at plot scale are commonly calculated based on either laser points or rasterized CHM cells, even if the same formulas as listed in Section 4.3 are used. The metrics based on points could be sensitive to the flight conditions and the sensor setting (Roussel et al. 2017). In contrast, CHM-based metrics can reduce such variations by focusing on only the canopy surface heights, which, however, miss the structural variations within canopy. Some researchers found that the different ways of generating metrics have small impacts on the performance of predicting forest attributes. For example, Lu et al. (2012) extracted four sets of metrics based on (1) all returns, (2) first returns, (3) last returns, and (4) CHM cells for 77 plots in mixed conifer forests in the Sierra Nevada in California. They found that the coefficient of determination (R^2) only slightly varied from 0.75 to 0.77 when the different sets of metrics were used to predict biomass. Chirici et al. (2016) compared point-based and CHM-based metrics to predict forest biomass over a study area of 36,360 ha with deciduous forests in central Italy. They found that the model-assisted estimates of forest biomass were similar for both sets of metrics, regardless of whether parametric or non-parametric methods were used.

Third, although numerous LiDAR metrics can be generated from a point cloud or CHM, the metrics used to predict forest attributes should be carefully chosen to both increase the accuracy and enhance the interpretability of the models (Chen 2013, Magnussen et al. 2016). Lu et al. (2016) summarized that LiDAR metrics can be classified based on whether they characterize (1) horizontal structure (2D), (2) vertical structure (1D), or (3) both horizontal and vertical (i.e., 3D) structure of canopy. If forest attributes of interest are the total or density value of each plot (e.g., basal area, stem volume, biomass, or LAI), 3D LiDAR metrics should be used. Examples of 3D LiDAR metrics include canopy geometric volume proposed by Chen et al. (2007). When LiDAR height metrics are calculated by including both canopy and ground returns, they are also essentially 3D metrics (Lu et al. 2016).

4.6 MODELING, MAPPING, AND ESTIMATING BIOMASS

The elevated concentration of carbon dioxide (CO_2), as a greenhouse gas, in the atmosphere is of major concern to our earth. Forests can absorb CO_2 in the atmosphere via photosynthesis and release O_2 to the atmosphere via respiration, the balance of which results in changes of forest biomass. Therefore, forest biomass is a key climate variable for the global carbon cycle and has attracted the attention of both scientists and policy makers.

However, forest biomass is also an attribute that is very difficult to estimate. It can be estimated using field measurements such as tree height, stem diameter, and wood density by applying allometric models obtained from destructive sampling and weighing of dried vegetation (Zolkos et al. 2013, Lu et al. 2016), and in situ measurements can be extended to larger areas using remote sensing methods (Frolking et al. 2009, Houghton et al. 2009). Conventional optical and radar imagery have the signal saturation problem in that the remotely sensed variables do not respond to biomass changes when biomass is relatively high (>100–200 Mg/ha); in contrast, LiDAR-derived height is strongly correlated with biomass even when biomass is as high as 1000 Mg/ha (Chen 2013). Therefore, LiDAR has emerged as the most promising technology for biomass estimation, especially over tropical forests when biomass density is high. Figure 4.20 shows some common LiDAR-derived and field-measured parameters that can be used as input to regression models for biomass estimation. A review on LiDAR data for biomass studies can be found in the works of Chen (2013) and Man et al. (2014). Table 4.4 summarizes some previous studies on LiDAR-based biomass modeling. Like other forest attributes, biomass can be estimated using individual-tree or area-based approaches from LiDAR. Most previous studies have focused on area-based approaches, which is also the focus of this section.

Area-based approaches require estimates of biomass over field plots to develop LiDAR-based models. Plot-level biomass is commonly calculated as the sum of the tree-level biomass divided by plot area (i.e., biomass density in units such as Mg/ha). The most accurate way of measuring tree biomass is to destructively harvest the trees and weigh them. However, such a method is impossible to apply at the landscape level. Instead, tree biomass in the field is estimated from other easily measurable variables (mainly DBH and height, and sometimes wood density) using allometric

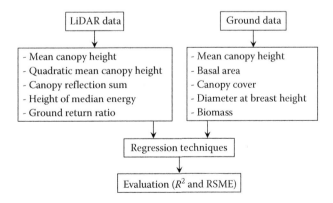

FIGURE 4.20   LiDAR-derived and field-measured parameters for forest biomass estimation.

TABLE 4.4

Summary of Previous Studies on LiDAR-Based Biomass Modeling

References	Parameters	Models	Results
Nelson et al. (1988)	Average of the three greatest laser heights, mean plot height (all pulses and canopy pulses), distance between the top of canopy and a point 2, 5 or 10 m above ground	Two logarithmic equations	$R^2 = 0.55$
Means et al. (1999)	LiDAR canopy height, quadratic mean canopy height, canopy reflectance sum	Allometric equations on DBH	$R^2 = 0.96$
Lefsky et al. (2002)	Max/min canopy height, canopy cover, variability in the upper canopy surface, total volumes of foliage and empty space in canopy	Stepwise multiple regression	$R^2 = 0.86$
Drake et al. (2002)	LiDAR canopy height, height of median energy, height /median ratio, ground return ratio	The tropical wet allometric equation	$R^2 = 0.93$
Nelson et al. (2004)	Quadratic mean height of pulses in the forest canopy	Parametric linear regression, nonparametric linear regression	$R^2 = 0.66$
Popescu et al. (2004)	Average/maximum crown diameter; maximum height	Regression models	$R^2 = 0.82$ (Pine), $R^2 = 0.33$ (hardwoods)
Zhao et al. (2009)	LiDAR-derived canopy height distributions canopy height quartile functions	A linear functional model and an equivalent nonlinear model	$R^2 = 0.95$
García et al. (2010)	LiDAR height, intensity or height combined with intensity data	A stepwise regression	$R^2 = 0.85$ (Pine), $R^2 = 0.70$ (Spanish juniper), $R^2 = 0.90$ (Holm oak)
Zhao et al. (2011)	LiDAR composite metrics	Support vector machine and Gaussian processes	RMSE = 21.4 (40.5) Mg/ha

Source: Man, Q. et al., *J. Appl. Remote Sens.*, 8, 081598, 2014.

models. Therefore, the very first and extremely important step in biomass estimation is to choose an appropriate allometric model to estimate field-based biomass. Chen (2015) compared three types of allometric models in the United States: one based on DBH only (the Jenkins method), one based on DBH and height [the Forest Inventory Analysis regional method], and another based on the expansion of stem volume to biomass [the Component Ratio Method (CRM)]. It was found that the use of different allometric models had substantial impact on the performance of LiDAR-based bio-mass models over three study sites in the Western United States. For example, over the forests in the Lake Tahoe basin in California, he found that the R^2 for one of the LiDAR-based biomass models varied dramatically from 0.51 to 0.92 when the allo-metric models were switched from the Jenkins method to CRM.

However, it should be noted that it is a wrong practice to choose the allome-tric model simply based on the maximum performance (e.g., as indicated by R^2) of LiDAR-based biomass models. The gold standard of choosing an allometric model is to compare the predictions of allometric models with destructive measurements of tree biomass at the study area (e.g., Colgan et al. 2014). Practically, this is rarely done because of the labor-intensive and expensive process of destructive measure-ments. Allometric models were usually developed using destructive measurements of trees at site(s) different from the site where the allometric models are applied to. The common criteria of choosing allometric models include the similarity of forest conditions between allometric model development site and application site, the num-ber of harvested trees, the range of tree size, and similarity of fitness (e.g., R^2 and root mean square error – RMSE) of the models. Overall, an analyst should choose an allometric model that was developed from a site with similar environmental condi-tions, with a large sample of harvest trees that spans a large range of tree size, and with similar fit to the model.

The goodness of fit of allometric models is closely related to the predictors that are used to predict biomass. The most common predictor is DBH, followed by tree height, and wood density. Such an order also reflects the degree of difficulty or ease of measuring corresponding variables in the field. From the perspective of model generality, allometric models that include these three predictors may have better pre-diction accuracy when they are applied beyond the sites where they were developed. For example, Chave et al. (2014) and Chen et al. (2015) fitted allometric models for over 4000 trees in the pan-tropics and found that the tree biomass has a nearly linear relationship with $\rho \times DBH^2 \times H$, where ρ is wood density and H is tree height. This indicates the importance of these three variables in predicting biomass.

Understanding allometry of tree biomass as previously discussed provides key insights in improving the modeling of biomass using LiDAR data. Note that LiDAR can provide direct information only for H, not for DBH and wood density. Although previous research has been done to predict DBH and wood density from LiDAR data, their estimation relies on their correlations with H. Therefore, to provide infor-mation about wood density and DBH independently from LiDAR-derived vegetation height, other data sources should be utilized. Since both the DBH-H relationship and wood density are dependent on species, the incorporation of information regard-ing tree species or, in general, vegetation type can potentially improve the biomass estimation. Along this line of thought, Chen et al. (2012) used mixed-effects mod-els to combine vegetation type information derived from aerial photographs with LiDAR metrics to estimate biomass. This is equivalent to developing vegetation-type-specific biomass models. It was found that such a combination can improve the biomass estimation for a conifer forest in Sierra Nevada, CA. A similar study was done by Chen et al. (2016) in Brazil, in which they stratified field plots over agrofor-estry plantations based on DBH-H relationship and wood density. They found that the mixed-effects modeling of biomass based on such a stratification can improve the model R^2 from 0.38 to 0.64 for their study site.

Although the analysis of many studies just focused on the results of LiDAR-based modeling of biomass over field plots, the biomass estimates over field plots by themselves are not very useful from an application perspective. This is because field plots only cover a tiny portion of forests over a study area. Forest managers and decision makers typically want to know the biomass over different locations, beyond the forest plots where trees are measured in the field. This requires "mapping" forest biomass over the whole study area. A user may also be interested in the mean or total biomass of the whole study area or a part of it. If we consider a study area as a geographic space tessellated with individual pixels, each pixel can be considered as a population unit and all pixels of the study year consist of the study population. From the perspective of sampling and statistics, the "estimation" of mean or total biomass of an area that consists of many pixels is a statistical inference process. Therefore, the specific applications of LiDAR-based biomass studies can be classified as either "mapping" biomass of individual pixels or "estimating" mean/total biomass of an area. Conventionally, geographers are interested in biomass mapping whereas foresters are interested in biomass estimation. Note that here we follow statisticians, for whom "estimation" specifically means the inference of population or subpopulation statistics from a probabilistic sample. However, the remote sensing literature usually uses the word "estimate" or "estimation" at many different scales including individual trees, plots, pixels, or a study area without referring to sampling and inference. We do not use strictly one single definition throughout the book. Instead, readers need to interpret the specific meaning according to the context of our discussion.

No matter if the application focus is biomass mapping or estimation, one big challenge is to characterize the errors of the estimated biomass at the pixel- or population/subpopulation-level. One very important factor that can affect the error analysis method is the sampling scheme of field plots, which could be probabilistic (e.g., simple random sampling (SRS) or stratified random sampling) or purposive (i.e., the location of field plots is determined on purpose to reduce the prediction errors and minimize the cost). Most LiDAR-based biomass studies used purposive field plots because the cost is too prohibitive and/or some locations might be inaccessible to collect probabilistic random samples. The error analysis of these studies has largely been limited to the calculation of model fitting statistics such as RMSE and R^2, sometimes using cross-validation methods. However, the model fitting statistics calculated from purposive samples, even with cross-validation, do not have strict statistical meaning (i.e., the sampling distribution of the fitting statistics is not Gaussian). Simply put, the fitting statistics of LiDAR-based biomass models based on purposive samples do not tell much information about the prediction performance of the same model applied to the population. More specifically, this means that an analyst may choose a purposive sample of field plots to develop a LiDAR-based model with very high R^2 and very low RMSE for the plots themselves, but the RMSE and R^2 could be much worse when the model is applied to the whole study area. So, the model fitting statistics from purposive samples are often misleading, if not useless, when the purpose is to infer the model performance for the whole study area.

A more rigorous method of error analysis is to characterize the error of model predictions for all pixels of the study area. Chen et al. (2015) introduced an uncertainty analysis method that can comprehensively characterize the errors in the whole process of LiDAR-based biomass modeling and prediction, including errors from field measurements, allometric modeling, tree-level biomass prediction, LiDAR data, and LiDAR-based biomass modeling and prediction. The method was applied to aboveground biomass (AGB) mapping in tropical forests in Ghana (Figure 4.21). The uncertainty analysis framework is useful in that (1) it allows mapping of biomass prediction uncertainty at the pixel level, (2) it can use both probabilistic and purposive samples, and (3) it is comprehensive and can help understand the error sources by quantifying the prediction errors caused by model parameters, predictors, and residuals, the three key components of a statistical model.

Chen et al. (2016) further expanded the above framework of uncertainty analysis from pixel level to any area that consists of multiple pixels (Figure 4.22). The method was applied to the biomass mapping and uncertainty analysis for an area of 69,508 km^2 in Northeastern Minnesota, USA. First, AGB was predicted at a pixel of 13 m resolution, which is equivalent to the forest plot size. Then, AGB was aggregated to coarse spatial resolutions. Maps of uncertainty can be produced at different spatial resolutions (Figure 4.23). Such a method differs from others in that it characterizes the biomass prediction errors for every location, not just for the locations of the field plots.

The previous uncertainty analysis method is one of the model-based approaches. If a probabilistic sample of field plots is used, the mean and uncertainty of biomass

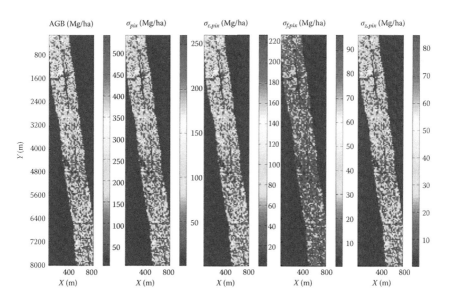

FIGURE 4.21 LiDAR-based maps of aboveground biomass and uncertainty. σ_{pix} is the pixel-level AGB error, $\sigma_{\varepsilon,pix}$, $\sigma_{f,pix}$, and $\sigma_{z,pix}$ are errors related to LiDAR-biomass model residuals, model parameters, and LiDAR metrics, respectively. (From Chen, Q. et al., *Remote Sens. Environ.*, 160, 134–143, 2015.)

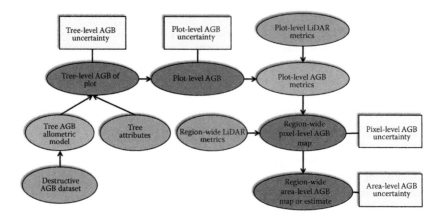

FIGURE 4.22 Framework of error propagation for the uncertainty analysis of biomass predictions. (Adapted from Chen, Q. et al., *Remote Sens.*, 8, 21, doi: 10.3390/rs8010021, 2016.)

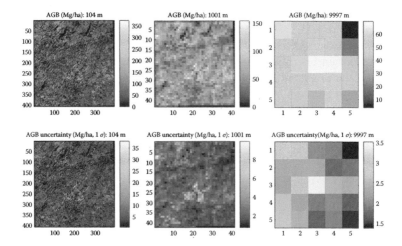

FIGURE 4.23 Maps of AGB mean and uncertainty of the same area but at different spatial resolutions. (Adapted from Chen, Q. et al., *Remote Sens.*, 8, 21, doi: 10.3390/rs8010021, 2016.)

over the study area can be estimated using classical frequentist inference. Further, LiDAR-based biomass maps can be combined with probabilistic samples to reduce the uncertainty of the estimator of mean biomass via the model-assisted method (McRoberts 2010, Næsset et al. 2011). Note that the conventional model-assisted method requires wall-to-wall LiDAR-based biomass maps to estimate the population mean and uncertainty.

At the regional scales (over tens of thousands of square kilometers or larger), few studies have processed large amount of wall-to-wall airborne LiDAR data to apply model-based or model-assisted approaches for uncertainty analysis (see Chen et al.

2016 for an exception). It is not very cost-effective to use wall-to-wall LiDAR data if the purpose is to estimate the biomass and uncertainty not for individual pixels, but for the areal mean. For estimating areal mean biomass, a better strategy is to use samples of LiDAR data in combination with field plots, sometimes with wall-to-wall optical or radar imagery, to improve the estimates. Along this line, many uncertainty analysis methods have been developed (Ståhl et al. 2010, Gregoire et al. 2010, Næsset et al. 2011, Gobakken et al. 2012).

PROJECT 4.1: EXTRACTING CANOPY HEIGHTS FROM LEAF-ON AND LEAF-OFF LiDAR DATA IN SUSQUEHANNA SHALE HILLS, PA, USA

1. Introduction

   The physical structure of trees derived from LiDAR data can be affected by seasonal changes (leaf-on and leaf-off) in canopy condition. For example, a leaf-on dense canopy may have concentrated LiDAR returns near the top of the canopy, with fewer returns from understory vegetation and the ground floor. In this project, leaf-on and leaf-off LiDAR data are separately used for canopy height extraction in a 1 km × 1 km study area in Susquehanna Shale Hills, PA, USA. CHMs are obtained by subtracting a digital terrain model (DTM) from a digital surface model (DSM). Since no field data is available for quantitative assessment, the results are only used for visual comparison. The objective is to understand that CHMs derived from leaf-on and leaf-off LiDAR data are usually different, and that care must be taken when applying leaf-on models to leaf-off LiDAR data for estimating forest inventory attributes, and vice versa.

2. Data

   The LiDAR data was collected during peak leaf-on (July 2010) and leaf-off (December 2010) conditions. The data in the 1 km × 1 km study area is based on data services provided by the OpenTopography facility with support from the National Science Foundation under NSF Award Numbers 0930731 and 0930643. The LiDAR point density is 13.59 points/m^2 (leaf-on) and 15.52 points/m^2 (leaf-off). The horizontal coordinate system is UTM Z18N NAD83 (CORS96) [EPSG: 26918], and the vertical coordinate system NAVD88 (Geoid 03) [EPSG: 5703]. LiDAR data acquisition and processing were completed by the National Center for Airborne Laser Mapping (NCALM), funded by the National Science Foundation Award EAR-0922307, and coordinated by Qinghua Guo for the Susquehanna Shale Hills Critical Zone Observatory funded by the National Science Foundation Award EAR-0725019. LiDAR point clouds in LAS format "leaf-on.las" (Figure 4.24) and "leaf-off.las" for the study area can be downloaded (by right-clicking each file and saving it to a local folder) from: http://geography.unt.edu/~pdong/LiDAR/Chapter4/Project4.1/

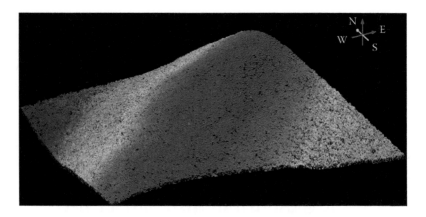

FIGURE 4.24   Leaf-on LiDAR point clouds in the 1 km × 1 km study area.

3. Project Steps

1. Open an empty Word document so that you can copy any results from the following steps to the document. To copy the whole screen to your Word document, press the PrtSc (print screen) key on your keyboard, then open your Word document and click the "Paste" button or press Ctrl+V to paste the content into your document. To copy an active window to your Word document, press Alt+PrtSc, then paste the content into your document.

2. Open ArcMap, and turn on the Spatial Analyst Extension and 3D Analyst Extension.

3. Create a LAS dataset. Go to ArcToolbox → Data Management Tools → LAS Dataset → Create LAS Dataset, and use leaf-on.las as input to create a LAS dataset leaf-on.lasd.

4. Create leaf-on DTM. Open the properties form of leaf-on.lasd and select the Filter panel. Select class "Ground" under "Classification Codes" in the Filter panel which means only ground points will be used for subsequent conversion and then click OK. Open ArcToolbox → Conversions → LAS Dataset to Raster, select "leaf-on.lasd" from the drop-down list as the input LAS dataset, "leafondtm" as the output raster in the output folder, ELEVATION as the field value, Binning as the interpolation type, AVERAGE as the cell assignment type, NATURAL_ NEIGHBOR as the void fill method, FLOAT as the output data type, CELLSIZE as the sampling type, 1 as the sampling value, and 1 as the Z factor; then click OK to create the output DTM raster (Figure 4.25). Note: You should select the input LAS dataset from the drop-down list because the filter was defined through the layer properties form. If you use the browse button to select a LAS dataset as input, all the data points in the LAS files it references will be processed, and the defined filter will not be used.

FIGURE 4.25 Output DTM converted from leaf-on LAS dataset.

5. Create leaf-on DSM. Open the properties form for leaf-on.lasd and select the Filter panel. Select "All Classes" under "Classification Codes" in the Filter panel, then select "Return 1" (First Return) under "Returns", then click OK. Use the same process as Step 4 to create the leaf-on DSM "leafondsm" (Figure 4.26).

6. Create leaf-on CHM. The leaf-on canopy height model is obtained by subtracting DTM from DSM in the Raster Calculator, which can be launched from ArcToolbox → Spatial analyst Tools → Map Algebra → Raster Calculator. Use "leafondsm" - "leafondtm" as the expression, "leafonchm" as the output, click OK to create the leaf-on CHM "leafonchm" (Figure 4.27).

7. Create leaf-off CHM. Repeat Steps 3–6 using leaf-off LiDAR data to create leaf-off DTM (which should be very close to leaf-on DTM because topography does not change very much between two LiDAR acquisition dates), leaf-off DSM (Figure 4.28), and leaf-off CHM (Figure 4.29).

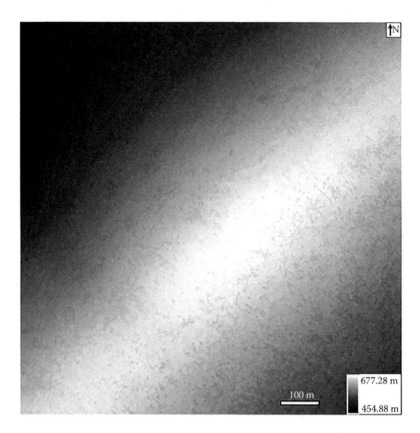

FIGURE 4.26 Output DSM converted from leaf-on LAS dataset.

8. Visual comparison of leaf-on and leaf-off CHM using the LAS Dataset
   Toolbar. In addition to visual comparison of the CHM spatial patterns
   in Figures 4.27 and 4.29, you can compare the leaf-on and leaf-off point
   clouds using the LAS Dataset Profile View tool on the LAS Dataset
   Toolbar as explained in Project 2.2.

9. Visual comparison of leaf-on and leaf-off CHM using the 3D Analyst
   Toolbar. The LAS Profile View tool in Step 8 provides profile views but
   does not include height and distance values. For better comparison of the
   CHMs, you can open the Customize menu of ArcMap and select Toolbars
   → 3D Analyst to open the 3D Analyst toolbar. Select the leaf-on CHM as
   the 3D Analyst Layer on the toolbar, then click the Interpolate Line tool,
   and draw a straight line from the northwest (NW) corner to the southeast
   (SE) of the CHM (using double click to end the line). Now click the Profile
   Graph icon on the 3D Analyst Toolbar to display the NW-SE profile for the
   leaf-on CHM (Figure 4.30 top). Select the leaf-off CHM as the 3D Analyst
   Layer and repeat the above process to create the NW-SE profile for the
   leaf-off CHM (Figure 4.30 bottom). Using the same process, you can create the NE-SW profiles for leaf-on and leaf-off CHMs (Figure 4.31).

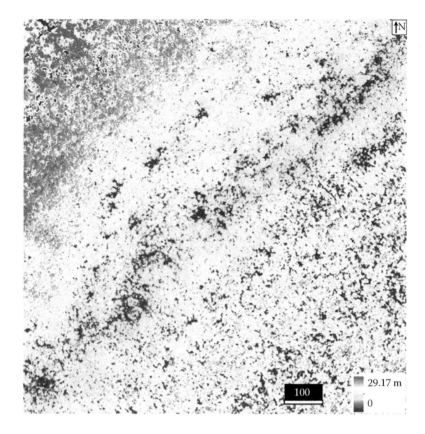

FIGURE 4.27   Leaf-on canopy height model.

10. Save your ArcMap project.
11. Questions: (1) What are the major differences between the leaf-on and leaf-off CHM profiles in Figures 4.30 and 4.31? (2) Suppose you created a model to estimate DBH of individual trees using tree heights derived from leaf-off LiDAR data; can you apply the model to leaf-on LiDAR data? Why or why not?

PROJECT 4.2:   IDENTIFYING DISTURBANCES FROM HURRICANES AND LIGHTNING STRIKES TO MANGROVE FORESTS USING LiDAR DATA IN EVERGLADES NATIONAL PARK, FL, USA

1. Introduction

Hurricanes and lighting strikes have impacted the mangrove forests along the coasts of Everglades National Park (ENP) in Florida, USA, producing gaps by causing the death of trees. To better understand the biological and

FIGURE 4.28 Output DSM converted from leaf-off LAS dataset.

environmental effects of these disturbances, a census of gaps at broader spatial scales is needed. Field plot-survey methods can be difficult and expensive, whereas optical remote sensing methods such as aerial photography may not be effective due to the lack of tree height information. Zhang et al. (2008) used LiDAR measurements derived before and after Hurricanes Katrina and Wilma (2005) to evaluate the impact of hurricanes and lighting strikes on the mangrove forests in ENP, and found that the proportion of high tree canopy decreased significantly after the 2005 hurricane season. In this project, LiDAR data acquired in ENP in November 2012 is used to demonstrate a complete process in ArcGIS for identifying and mapping disturbances from hurricanes and lightning strikes to mangrove forests.

2. Data

LiDAR point clouds with a point density of 7.95 points/m^2 from a 1 km × 1 km study area near Shark River in ENP is used in this project. The horizontal coordinate system is UTM Z17N NAD83 (2011) [EPSG: 26917], and the vertical coordinate system NAVD88 (Geoid 12A) [EPSG: 5703]. LiDAR data acquisition and processing was completed by the National Center for Airborne Laser Mapping (NCALM—http://www.ncalm.org). NCALM funding was provided by the National Science Foundation (NSF) Division

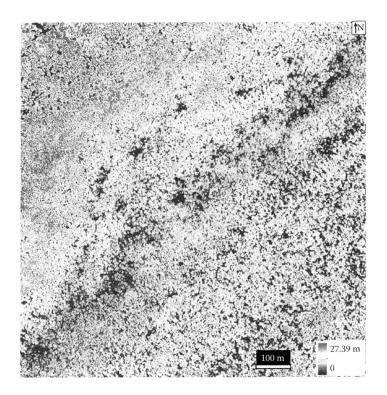

FIGURE 4.29   Leaf-off canopy height model.

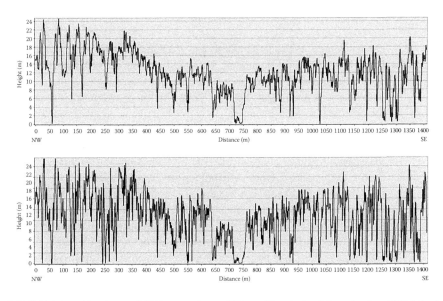

FIGURE 4.30   Northwest (NW) to southeast (SE) profiles derived from leaf-on CHM (top) and leaf-off CHM (bottom).

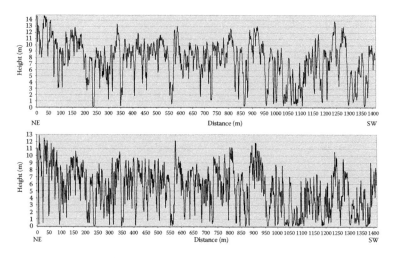

FIGURE 4.31 Northeast (NE) to southwest (SW) profiles derived from leaf-on CHM (top) and leaf-off CHM (bottom).

of Earth Sciences, Instrumentation and Facilities Program, EAR-1043051. This dataset is based on data services provided by the OpenTopography facility with support from the NSF Award Numbers 0930731 and 0930643. LiDAR data file "Mangrove.las" in LAS format can be downloaded (by right-clicking the file and saving it to a local folder) from: http://geography. unt.edu/~pdong/LiDAR/Chapter4/Project4.2/

3. Project Steps

   1. Open an empty Word document so that you can copy any results from the following steps to the document. To copy the whole screen to your Word document, press the PrtSc (print screen) key on your keyboard, then open your Word document and click the "Paste" button or press Ctrl+V to paste the content into your document. To copy an active window to your Word document, press Alt+PrtSc, then paste the content into your document.
   2. Open ArcMap, go to the Customize menu and select "Extensions...". Check the Spatial Analyst Extension, 3D Analyst Extension, and ArcScan.
   3. Create LAS dataset. Open ArcToolbox → Data Management Tools → LAS Dataset → Create LAS Dataset. Use Mangrove.las as input and Mangrove.lasd as output to create a LiDAR dataset. The LAS dataset is added to ArcMap automatically. Note: You should use an output folder that you have full control in this project. Do not use default output folders because they may be protected and you may not be allowed to edit files in the protected folders.
   4. Select ground points. Open the Properties form of Mangrove.lasd in ArcMap, select the Filter tab, select "Ground" under "Classification Codes", then click OK.

5. Create DTM. Open ArcToolbox → Conversion Tools → To Raster → LAS Dataset to Raster. Select "Mangrove" from the drop-down list as the input LAS dataset, "dtm" as the output raster in the output folder, ELEVATION as the field value, Binning as the interpolation type, AVERAGE as the cell assignment type, NATURAL_NEIGHBOR as the void fill method, FLOAT as the output data type, CELLSIZE as the sampling type, 1 as the sampling value, and 1 as the Z factor, then click OK to create the output DTM raster. Note: You should select the input LAS dataset from the drop-down list because the filter was defined through the layer properties form in Step 4. If you use the browse button to select a LAS dataset as input, all the data points in the LAS files it references will be processed, and the defined filter will not be used.

6. Create DSM. Open the Properties form of Mangrove.lasd in ArcMap, select the Filter tab, select "All Classes" under "Classification Codes", select "Return 1" (First Return) under "Returns", then click OK. Use "dsm" as the output raster in Step 5 to create the output DSM raster.

7. Create CHM. Open ArcToolbox → Spatial Analyst Tools → Map Algebra → Raster Calculator, use "dsm" - "dtm" as the expression and "chm" as the output raster in the output folder to create the CHM (Figure 4.32).

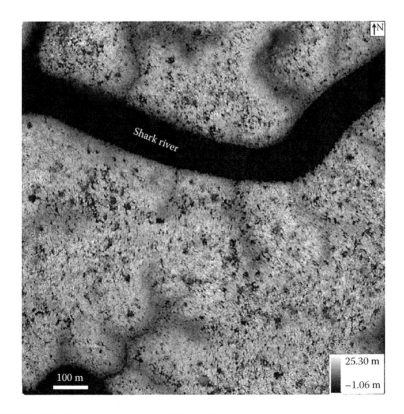

FIGURE 4.32   Mangrove canopy height model.

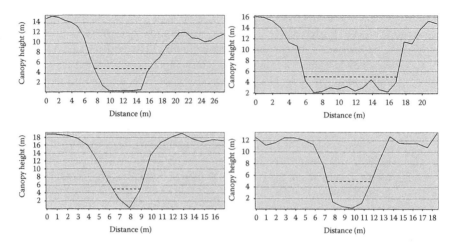

FIGURE 4.33 Sample profiles for gaps in CHM.

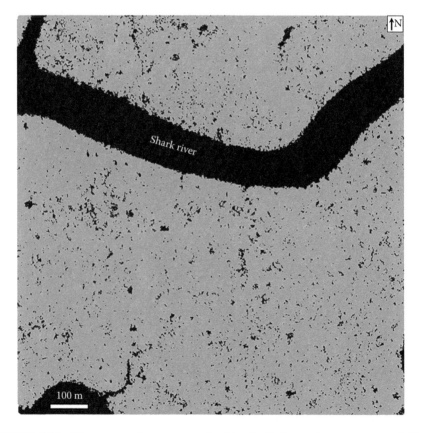

FIGURE 4.34 Binary raster for two categories: blue for 1's (gaps, river segments, and tidal creeks), and green for 0's (canopy over 5 m in height).

FIGURE 4.35   ArcScan Toolbar in ArcMap.

Select Connected Cells   X

Choose the type of connected cell selection:

Total area of connected cells   ▼

Choose where to search for connected cells:

Foreground   ▼

Choose the operator to constrain the selection:

less than or equal to   ▼

Enter total area: 4   raster pixels

Choose how the result will affect the current raster selection:

Create a new selection   ▼

☐ Select only from the current extent.

                              OK   Cancel

FIGURE 4.36   Selecting connected cells using ArcScan Toolbar.

8. Examine sample gap profiles. Zoom in to the gaps (dark holes in Figure 4.32), then open the "Customize" menu of ArcMap and select Toolbars → 3D Analyst. Select the canopy height model "chm" as the 3D Analyst Layer for the 3D Analyst toolbar, then use the "Interpolate Line" tool on the 3D Analyst toolbar to draw a straight line across a gap, and click the "Profile Graph" button on the toolbar to show the CHM profile. Repeat this process several times to determine a proper height threshold that can be used to extract possible gaps. Figure 4.33 shows four sample profiles for gaps in the CHM, and a height threshold of 5 m (manually added dashed lines in Figure 4.33) is used to separate possible gaps from the CHM. Although this threshold is purely empirical, slight variations in the height threshold (for example, 6 m or 7 m) usually do not significantly change the size of the gaps because many gaps in the CHM have steep slopes on the edges, as can be seen in Figure 4.33.

9. Create a binary raster using the height threshold (5 m). Open ArcToolbox → Spatial Analyst Tools → Map Algebra → Raster Calculator, use Con("chm" <= 5, 1, 0) as the expression to create a binary output raster "binary_gaps" in the output folder (Figure 4.34), where blue cells (1's) are for gaps, river segments, and tidal creeks; and green cells (0's) for canopy over 5 m in height. The noises, river segments, and tidal creeks will be removed in the following steps.

10. Edit gaps in binary raster. Open the "Customize" menu of ArcMap and select Toolbars → Editor. Right-click the binary raster layer "binary_gaps" in the ArcMap table of contents, then select Editing Features → Start Editing. Then open the Customize menu in ArcMap and select Toolbars → ArcScan. Select "binary_gaps" as the ArcScan Raster Layer for the ArcScan Toolbar (Figure 4.35).

11. Open the Cell Selection menu on the ArcScan Toolbar and select "Select Connected Cells…". Specify that the total area of connected cells in "Foreground" (blue cells) is less than or equal to 4 cells as shown in Figure 4.36, and click OK. The selected cells are shown as cyan cells in the sample window in Figure 4.37A. Now open the Raster Cleanup menu on the ArcScan Toolbar and select "Start Cleanup" to enable the tools under the Raster Cleanup menu. Click "Erase Selected Cells" in the Raster Cleanup menu to erase the selected cells (Figure 4.37B).

12. From Figure 4.37B, it can be seen that some gaps have isolated cells (noises or isolated trees). These cells can be selected using the same process as in Step 11 by choosing "Background" when searching for connected cells (see Figure 4.36). Once the background cells are selected (Figure 4.38A),

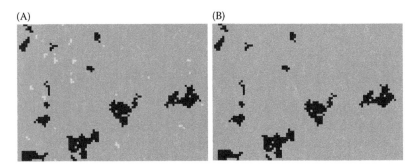

FIGURE 4.37 Removing selected foreground cells. (A) Connected cells less than four pixels are selected and (B) Selected foreground cells are erased.

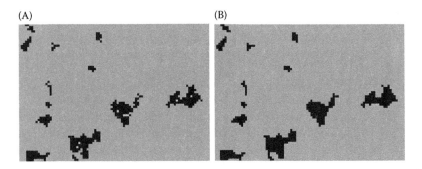

FIGURE 4.38 Filling selected background cells. (A) Connected cells less than four pixels are selected and (B) Selected background cells are filled with foreground color, then the closing operator is applied to the binary raster to clean the boundary.

select "Fill Selected Cells" under the Raster Cleanup menu to change the selected green cells to blue. Select "Closing…" under the Raster Cleanup menu, put 0.5 as the number of pixels, and click OK to get the output (Figure 4.38B). Note that the isolated cells and boundaries in Figure 4.38A can also be cleaned using neighborhood statistics in ArcGIS Spatial Analyst Tools (such as majority or median filters in windows of 3×3 cells). However, the neighborhood filters will also bring major changes to the boundaries of the cell groups. Therefore, a better option is to use the mathematical morphological operators (erosion, dilation, opening, and closing) under the Raster Cleanup menu of the ArcScan Toolbar.

13. Remove river segments and tidal creeks. The river segments and tidal creeks in Figure 4.34 should be removed so that the extracted gaps can be converted to a polygon shapefile. Use the "Select Connected Cells" tool as explained in Step 12 to select foreground cells (blue cells) that are greater than 600 raster pixels (Figure 4.39A), then select "Erase Selected Cells" under the Raster Cleanup menu to obtain the final gaps

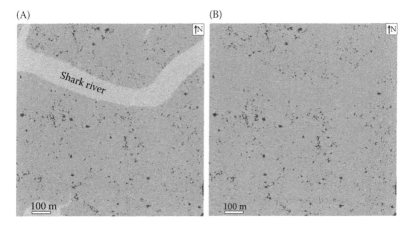

FIGURE 4.39 Erasing river segments and tidal creeks. (A) Connected cells are selected and (B) Selected cells are erased.

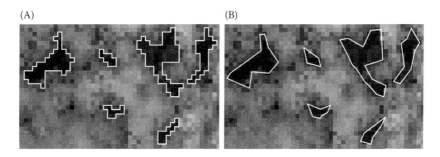

FIGURE 4.40 Sample area showing polygons for gaps in a mangrove forest. (A) original polygons extracted from LiDAR data and (B) simplified polygons.

FIGURE 4.41 A 1 m × 1 m gap polygon (lower right corner) connected to a larger gap polygon.

(Figure 4.39B). Click the Editor menu on the Editor Toolbar, and select "Stop Editing" to save the edits of "binary_gaps".

14. Convert binary raster to polygon shapefile. Open ArcToolbox → Conversion Tools → From Raster → Raster to Polygon. Use "binary_gaps" as the input raster, VALUE as the field, and "Mangrove_Gap1.shp" as the output polygon shapefile, uncheck the "Simplify polygon", then click OK to create the output polygon shapefile. Repeat the conversion process using "Mangrove_Gap2.shp" as the output, and check the "Simplify polygon" option to create output polygon shapefile with simplified boundaries. A GRIDCODE field is created in the attribute tables of output shapefiles where 1 is for gaps and 0 for background (canopy). Figure 4.40 is a comparison between the two output polygon shapefiles for mangrove gaps using the 1-m resolution CHM as the backdrop.

15. Save your ArcMap project and Word document.

16. Questions: (1) How do you remove the background polygons (GRIDCODE = 0)? (2) How do you remove the 1 m × 1 m gap polygons such as the one at the lower right corner of Figure 4.41? (3) How do you show the histogram of the gap sizes in the study area?

REFERENCES

Balenović, I., Alberti, G., and Marjanović, H., 2013. Airborne laser scanning—The status and perspectives for the application in the South-East European Forestry. *SEEFOR*, 4: 59–79.

Baltsavias, E.P., 1999. Airborne laser scanning: Basic relations and formulas. *ISPRS Journal of Photogrammetry and Remote Sensing*, 54: 199–214.

Ben-Arie, J.R., Hay, G.J., Power, R.P., Castilla, G., and St-Onye, B., 2009. Development of a pit canopy height models and airborne lasers to estimate forest biomass: Two problems. *International Journal of Remote Sensing*, 21: 2153–2162.

Breidenbach, J., Næsset, E., Lien, V., Gobakken, T., and Solberg, S., 2010. Prediction of species specific forest inventory attributes using a nonparametric semi-individual tree crown approach based on fused airborne laser scanning and multispectral data. *Remote Sensing of Environment*, 114: 911–924.

Cao, L., Coops, N.C., Innes, J.L., Dai, J.S., Ruan, H., and She, G., 2016. Tree species classification in subtropical forests using small-footprint full-waveform LiDAR data. *International Journal of Applied Earth Observation and Geoinformation*, 49: 39–51.

Chan, T. F., and Vese, L., 2001. Active contours without edges. *IEEE Transactions on Image Processing*, 10: 266–277.

Chave, J., Réjou-Méchain, M., Búrquez, A., Chidumayo, E., Colgan, M.S., Delitti, W.B., Duque, A., Eid, T., Fearnside, P.M., Goodman, R.C., and Henry, M., 2014. Improved allometric models to estimate the aboveground biomass of tropical trees. *Global Change Biology*, 20: 3177–3190.

Chen, Q., 2013. Lidar remote sensing of vegetation biomass. *Remote Sensing of Natural Resources* (edited by Q. Weng and G. Wang), 399–420.

Chen, Q., 2015. Modeling aboveground tree woody biomass using national-scale allometric methods and airborne lidar. *ISPRS Journal of Photogrammetry and Remote Sensing*, 106: 95–106.

Chen, Q., Baldocchi, D., Gong, P., and Kelly, M., 2006. Isolating individual trees in a savanna woodland using small footprint lidar data. *Photogrammetric Engineering and Remote Sensing*, 72: 923–932.

Chen, Q., Gong, P., Baldocchi, D. and Tian, Y.Q., 2007. Estimating basal area and stem volume for individual trees from lidar data. *Photogrammetric Engineering and Remote Sensing*, 73: 1355–1365.

Chen, Q., Vaglio Laurin, G., Battles, J.J., and Saah, D., 2012. Integration of airborne lidar and vegetation types derived from aerial photography for mapping aboveground live biomass. *Remote Sensing of Environment*, 121: 108–117.

Chen, Q., Vaglio Laurin, G., and Valentini, R., 2015. Uncertainty of remotely sensed aboveground biomass over an African tropical forest: Propagating errors from trees to plots to pixels. *Remote Sensing of Environment*, 160: 134–143.

Chen, Q., Lu, D., Keller, M., dos-Santos, M.N., Bolfe, E.L., Feng, Y., and Wang, C., 2016. Modeling and mapping agroforestry aboveground biomass in the Brazilian Amazon using airborne lidar data. *Remote Sensing*, 8(1): 21, doi: 10.3390/rs8010021.

Chen, Q., McRoberts, R.E., Wang, C., and Radtke, P.J., 2016. Forest aboveground biomass mapping and estimation across multiple spatial scales using model-based inference. *Remote Sensing of Environment*, 184: 350–360.

Chirici, G., McRoberts, R.E., Fattorini, L., Mura, M., and Marchetti, M., 2016. Comparing echo-based and canopy height model-based metrics for enhancing estimation of forest aboveground biomass in a model-assisted framework. *Remote Sensing of Environment*, 174: 1–9.

Chopping, M., 2011. CANAPI: Canopy analysis with panchromatic imagery. *Remote Sensing Letters*, 2: 21–29.

Clark, M.L., Roberts, D.A., Ewel, J.J., and Clark, D.B., 2011. Estimation of tropical rain forest aboveground biomass with small-footprint lidar and hyperspectral sensors. *Remote Sensing of Environment*, 115: 2931–2942.

Colgan, M.S., Swemmer, T., and Asner, G.P., 2014. Structural relationships between form factor, wood density, and biomass in African savanna woodlands. *Trees*, 28: 91–102.

Dong, P., 2009. Characterization of individual tree crowns using three-dimensional signatures derived from LiDAR data. *International Journal of Remote Sensing*, 30: 6621–6628.

Dong, P., 2010. Sensitivity of LiDAR-derived three-dimensional shape signatures for individual tree crowns: a simulation study. *Remote Sensing Letters*, 1: 159–167.

Drake, J.B., Dubayah, R.O., Knox, R.G., Clark, D.B., and Blair, J.B., 2002. Sensitivity of large-footprint LiDAR to canopy structure and biomass in a neotropical rainforest. *Remote Sensing of Environment*, 81: 378–392.

Duncanson, L.I., Niemann, K.O., and Wulder, M.A., 2010. Estimating forest canopy height and terrain relief from GLAS waveform metrics. *Remote Sensing of Environment*, 114: 138–154.

Duncanson, L.I., Cook, B.D., Hurtt, G.C., and Dubayah, R.O., 2014. An efficient, multi-layered crown delineation algorithm for mapping individual tree structure across multiple ecosystems. *Remote Sensing of Environment*, 154: 378–386.

Erdody, T.L., and Moskal, L.M., 2010. Fusion of LiDAR and imagery for estimating forest canopy fuels. *Remote Sensing of Environment*, 114: 725–737.

Evans, J.S., Hudak, A.T., Faux, R., and Smith, A.M.S., 2009. Discrete return Lidar in natural resources: Recommendations for project planning, data processing, and deliverables. *Remote Sensing*, 1: 776–794.

Ferraz, A., Bretar, F., Jacquemoud, S., Gonçalves, G., Pereira, L., Tomé, M., and Soares, P., 2012. 3-D mapping of a multi-layered Mediterranean forest using ALS data. *Remote Sensing of Environment*, 121: 210–223.

Fieber, K.D., Davenport, I.J., Tanase, M.A., Ferryman, J.M., Gurney, R.J., Becerra, V.M., Walker, J.P., and Hacker, J.M., 2015. Validation of canopy height profile methodology for small-footprint full-waveform airborne LiDAR data in a discontinuous canopy environment. *ISPRS Journal of Photogrammetry and Remote Sensing*, 104: 144–157.

Frazer, G.W., Magnussen, S., Wulder, M.A., and Niemann, K.O., 2011. Simulated impact of sample plot size and co-registration error on the accuracy and uncertainty of LiDAR-derived estimates of forest stand biomass. *Remote Sensing of Environment*, 115: 636–649.

Frolking, S., Palace, M.W., Clark, D.B., Chambers, J.Q., Shugart, H.H., and Hurtt, G.C., 2009. Forest disturbance and recovery: A general review in the context of spaceborne remote sensing of impacts on aboveground biomass and canopy structure. *Journal of Geophysical Research: Biogeosciences*, 114: G00E02.

García, M., Riaño, D., Chuvieco, E., and Danson, F.M., 2010. Estimating biomass carbon stocks for a Mediterranean forest in central Spain using LiDAR height and intensity data. *Remote Sensing of Environment*, 114: 816–830.

Gaveau, D.L.A., and Hill, R.A., 2003. Quantifying canopy height underestimation by laser pulse penetration in small-footprint airborne laser scanning data. *Canadian Journal of Remote Sensing*, 29: 650–657.

Gobakken, T., Næsset, E., Nelson, R., Bollandsås, O.M., Gregoire, T.G., Ståhl, G., Holm, S., Ørka, H.O., and Astrup, R., 2012. Estimating biomass in Hedmark County, Norway using national forest inventory field plots and airborne laser scanning. *Remote Sensing of Environment*, 123: 443–456.

Gregoire, T.G., Ståhl, G., Næsset, E., Gobakken, T., Nelson, R., and Holm, S., 2010. Model-assisted estimation of biomass in a LiDAR sample survey in Hedmark County, Norway. *Canadian Journal of Forest Research*, 41(1): 83–95.

Holmgren, J., Nilsson, M., and Olsson, H., 2003. Estimation of tree height and stem volume on plots using airborne laser scanning. *Forest Science*, 49: 419–428.

Houghton, R.A., Hall, F., and Goetz, S.J., 2009. Importance of biomass in the global carbon cycle. *Journal of Geophysical Research*, 114: G00E03.

Hu, B., Li, J., Jing, L., and Judah, A., 2014. Improving the efficiency and accuracy of individual tree crown delineation from high-density LiDAR data. *International Journal of Applied Earth Observation and Geoinformation*, 26: 145–155.

Hudak, A.T., Crookston, N.L., Evans, J.S., Falkowski, M.J., Smith, A.M.S., Morgan, P., and Gessler, P., 2006. Regression modeling and mapping of coniferous forest basal area and tree density from discrete-return lidar and multispectral satellite data. *Canadian Journal of Remote Sensing*, 32: 126–138.

Hudak, A.T., Crookston, N.L., Evans, J.S., Hall, D.E., and Falkowski, M.J., 2008. Nearest neighbor imputation modeling of species-level, plot-scale structural attributes from lidar data. *Remote Sensing of Environment*, 112: 2232–2245.

Hyde, P., Dubayah, R., Walker, W., Blair, J.B., Hofton, M., and Hunsaker, C., 2006. Mapping forest structure for wildlife habitat analysis using multi-sensor (LiDAR, SAR/InSAR, ETM+, Quickbird) synergy. *Remote Sensing of Environment*, 102: 63–73.

Hyyppä, J., and Inkinen, M., 1999. Detecting and estimating attributes for single trees using laser scanner. *The Photogrammetric Journal of Finland*, 16: 27–42.

Hyyppä, J., Pyysalo, U., Hyyppä, H., and Samberg, A., 2000. Elevation accuracy of laser scanning-derived digital terrain and target models in forest environment. *EARSel Proceedings LiDAR Workshop*, Dresden/FRG, volume 1, pp. 139–147.

Hyyppä, J., Kelle, O., Lehikoinen, M., and Inkinen, M., 2002. A segmentation based method to retrieve stem volume estimates from 3-D tree height models produced by laser scanners. *IEEE Transactions on Geoscience and Remote Sensing*, 39: 969–975.

Hyyppä, J., Hyyppä, H., Yu, X., Kaartinen, H., Kukko, A., and Holopainen, M., 2008. Forest inventory using small-footprint airborne LiDAR. In: (J. Shan, and C.K. Toth, eds) *Topographic Laser Ranging and Scanning: Principles and Processing*, CRC Press, Boca Raton, FL, pp. 335–370.

Jensen, J.L.R., Humes, K.S., Vierling, L.A., and Hudak, A.T., 2008. Discrete return lidar-based prediction of leaf area index in two conifer forests. *Remote Sensing of Environment*, 112: 3947–3957.

Jones, T.J., Coops, N.C., and Sharma, T., 2010. Assessing the utility of airborne hyperspectral and LiDAR data for species distribution mapping in the coastal Pacific Northwest, Canada. *Remote Sensing of Environment*, 114: 2841–2852.

Kaartinen, H., Hyyppä, J., Yu, X., Vastaranta, M., Hyyppä, H., Kukko, A., Holopainen, M., Heipke, C., Hirschmugl, M., Morsdorf, F., and Næsset, E., 2012. An international comparison of individual tree detection and extraction using airborne laser scanning. *Remote Sensing*, 4: 950–974.

Kato, A., Moskal, L.M., Schiess, P., Swanson, M.E., Calhoun, D., and Stuetzle, W., 2009. Capturing tree crown formation through implicit surface reconstruction using airborne lidar data. *Remote Sensing of Environment*, 113: 1148–1162.

Koch, B., Heyder, U., and Weinacker, H., 2003. Detection of individual tree crowns in airborne LiDAR data. *Photogrammetric Engineering and Remote Sensing*, 72: 357–363.

Kwak, D., Lee, W., Lee, J., Biging, G., and Gong, P., 2007. Detection of individual trees and estimation of tree height using LiDAR data. *Journal of Forest Research*, 12: 425–434.

Latifi, H., Fassnacht, F.E., Müller, J., Tharani, A., Dech, S., and Heurich, M., 2015. Forest inventories by LiDAR data: A comparison of single tree segmentation and metric-based methods for inventories of a heterogeneous temperate forest. *International Journal of Applied Earth Observation and Geoinformation*, 42: 162–174.

Leckie, D.G., Yuan, X., Ostaff, D.P., Piene, H., and MacLean, D.A., 1992. Analysis of high resolution multispectral MEIS imagery for spruce budworm damage assessment on a single tree basis. *Remote Sensing of Environment*, 40: 125–136.

Leckie, D., Gougeon, F., Hill, D., Quinn, R., Armstrong, L., and Shreenan, R., 2003. Combined high-density LiDAR and multispectral imagery for individual tree crown analysis. *Canadian Journal for Remote Sensing*, 29: 633–649.

Lefsky, M.A., 2010. A global forest canopy height map from the Moderate Resolution Imaging Spectroradiometer and the Geoscience Laser Altimeter System. *Geophysical Research Letters*, 37: L15401.

Lefsky, M.A., Cohen, W.B., Harding, D.J., Parker, G.G., Acker, S.A., and Gower, S.T., 2002. LiDAR remote sensing of above-ground biomass in three biomes. *Global Ecology and Biogeography*, 11: 393–399.

Lim, K., Treitz, P., Wulder, M., St-Onge, B., and Flood, M., 2003. LiDAR remote sensing of forest structure. *Progress in Physical Geography*, 27: 88–106.

Liu, H., and Dong, P., 2014. A new method for generating canopy height models from discrete-return LiDAR point clouds. *Remote Sensing Letters*, 5: 575–582.

Liu, T., Im, J., and Quackenbush, L.J., 2015. A novel transferable individual tree crown delineation model based on Fishing Net Dragging and boundary classification. *ISPRS Journal of Photogrammetry and Remote Sensing*, 110: 34–47.

Li, W., Guo, Q., Jakubowski, M.K., and Kelly, M., 2012. A new method for segmenting individual trees from the lidar point cloud. *Photogrammetric Engineering and Remote Sensing*, 78: 75–84.

Lu, D., Chen, Q., Wang, G., Moran, E., Batistella, M., Zhang, M., Vaglio Laurin, G., and Saah, D., 2012. Aboveground forest biomass estimation with Landsat and LiDAR data and uncertainty analysis of the estimates. *International Journal of Forestry Research*, Article ID 436537, doi: 10.1155/2012/436537.

Lu, D., Chen, Q., Wang, G., Liu, L., Li, G., and Moran, E., 2016. A survey of remote sensing-based aboveground biomass estimation methods in forest ecosystems. *International Journal of Digital Earth*, 9: 63–105.

MacMillan, R.A., Martin, T.C., Earle, T.J., and McNabb, D.J., 2003. Automated analysis and classification of landforms using high-resolution digital elevation data: Applications and issues. *Canadian Journal of Remote Sensing*, 29: 357–366.

Magnussen, S., Næsset, E., Kändler, G., Adler, P., Renaud, J.P., and Gobakken, T., 2016. A functional regression model for inventories supported by aerial laser scanner data or photogrammetric point clouds. *Remote Sensing of Environment*, 184: 496–505.

Maltamo, M., Mustonen, K., Hyyppä, J., Pitkanen, J., and Yu, X., 2004. The accuracy of estimating individual tree variables with airborne laser scanning in a boreal nature reserve. *Canadian Journal of Forest Research*, 34: 1791–1801.

Mallet, C., and Bretar, F., 2009. Full-waveform topographic LiDAR: State-of-the-art. *ISPRS Journal of Photogrammetry and Remote Sensing*, 64: 1–16.

Man, Q., Dong, P., Guo, H.D., Liu, G., and Shi, R., 2014. LiDAR and hyperspectral data for estimation of forest biomass: A review. *Journal of Applied Remote Sensing*, 8(1): 081598. doi: 10.1117/1.JRS.8.081598.

McRoberts, R.E., 2010. Probability-and model-based approaches to inference for proportion forest using satellite imagery as ancillary data. *Remote Sensing of Environment*, 114: 1017–1025.

Means, J.E., Acker, S.A., Harding, D.J., Blair, J.B., and Lefsky, M.A., 1999. Use of large-footprint scanning airborne LiDAR to estimate forest stand characteristics in the Western Cascades of Oregon. *Remote Sensing of Environment*, 67: 298–308.

Means, J.E., Acker, S.A., Brandon, J., Fritt, B.J., Renslow, M., Emerson, L., and Hendrix, C., 2000. Predicting forest stand characteristics with airborne scanning lidar. *Photogrammetric Engineering and Remote Sensing*, 66: 1367–1371.

Mei, C., and Durrieu, S., 2004. Tree crown delineation from digital elevation models and high resolution imagery. Laser-Scanners for Forest and Landscape Assessment, October 3–6, 2004, Freiburg, Germany.

Mongus, D., and Žalik, B., 2015. An efficient approach to 3D single tree-crown delineation in LiDAR data. *ISPRS Journal of Photogrammetry and Remote Sensing*, 108: 219–233.

Monnet, J.M., Mermin, È., Chanussot, J., and Berger, F., 2010. Tree top detection using local maxima filtering: A parameter sensitivity analysis. *Silvilaser 2010*, 14–17 September, Freiburg, Germany.

Morsdorf, F., Meier, E., Allgöwer, B., and Nüesch, D., 2003. Clustering in airborne laser scanning raw data for segmentation of single trees. *International Archives of the Photogrammetry, Remote Sensing and Spatial Information Sciences*, 34: 27–33.

Næsset, E., 1997. Determination of mean tree height of forest stands using airborne laser scanner data. *ISPRS Journal of Photogrammetry and Remote Sensing*, 52: 49–56.

Næsset, E., Gobakken, T., Solberg, S., Gregoire, T.G., Nelson, R., Ståhl, G., and Weydahl, D., 2011. Model-assisted regional forest biomass estimation using LiDAR and InSAR as auxiliary data: a case study from a boreal forest area. *Remote Sensing of Environment*, 115: 3599–3614.

Nelson, R., Krabill, W., and Tonelli, J., 1988. Estimating forest biomass and volume using airborne laser data. *Remote Sensing of Environment*, 24: 247–267.

Nelson, R., Jimenez, J., Schnell, C.E., Hartshorn, G.S., Gregoire, T.G., and Oderwald, R., 2000. Canopy height models and airborne lasers to estimate forest biomass: Two problems. *International Journal of Remote Sensing*, 21: 2153–2162.

Nelson, R., Short, A., and Valenti, M., 2004. Measuring biomass and carbon in Delaware using an airborne profiling LIDAR. *Scandinavian Journal of Forest Research*, 19: 500–511.

Nie, S., Wang, C., Dong, P., and Xi, X., 2016. Estimating leaf area index of maize using airborne full-waveform LiDAR data. *Remote Sensing Letters*, 7: 110–120.

Osada, R., Funkhouser, T., Chazelle, B., and Dobkin, D., 2002. Shape distribution. *ACM Transactions on Graphics*, 21: 807–832.

Persson, Å, Holmgren, J., and Söderman, U., 2002. Detecting and measuring individual trees using an airborne laser scanner. *Photogrammetric Engineering and Remote Sensing*, 68: 925–932.

Pitkanen, J., M. Maltamo, J. Hyyppa, and X. Yu. 2004. Adaptive methods for individual tree detection on airborne laser based canopy height model. *International Archives of Photogrammetry, Remote Sensing and Spatial Information Sciences*, XXXVI-8/W2: 187–191.

Popescu, S.C., and Wynne, R.H., 2004. Seeing the trees in the forest: using LiDAR and multispectral data fusion with local filtering and variable window size for estimating tree height. *Photogrammetric Engineering and Remote Sensing*, 70: 589–604.

Popescu, S.C., Wynne, R.H., and Nelson, R.F., 2003. Measuring individual tree crown diameter with lidar and assessing its influence on estimating forest volume and biomass. *Canadian Journal of Remote Sensing*, 29: 564–577.

Popescu, S.C., Wynne, R.H., and Scrivani, J.A., 2004. Fusion of small-footprint LiDAR and multispectral data to estimate plot-level volume and biomass in deciduous and pine forests in Virginia, USA. *Forest Science*, 50: 551–565.

Puttonen, E., Suomalainen, J., Hakala, T., Räikkönen, E., Kaartinen, H., Kaasalainen, S., and Litkey, P., 2010. Tree species classification from fused active hyperspectral reflectance and LIDAR measurements. *Forest Ecology and Management*, 260: 1843–1852.

Rahman, M.Z.A. and Gorte, B.G.H., 2009. Tree crown delineation from high resolution airborne lidar based on densities of high points. *International Archives of Photogrammetry, Remote Sensing and Spatial Information Sciences*, XXXVIII-3/W8: 123–128.

Reitberger, J., Schnorr, C., Krzystek, P., and Stilla, U., 2009. 3D segmentation of single trees exploiting full waveform LIDAR data. *ISPRS Journal of Photogrammetry and Remote Sensing*, 64: 561–574.

Roussel, J.R., Caspersen, J., Béland, M., Thomas, S., and Achim, A., 2017. Removing bias from LiDAR-based estimates of canopy height: Accounting for the effects of pulse density and footprint size. *Remote Sensing of Environment*, 198: 1–16.

Ståhl, G., Holm, S., Gregoire, T.G., Gobakken, T., Næsset, E., and Nelson, R., 2010. Model-based inference for biomass estimation in a LiDAR sample survey in Hedmark County, Norway. *Canadian Journal of Forest Research*, 41: 96–107.

Strîmbu, V.F., and Strîmbu, B.M., 2015. A graph-based segmentation algorithm for tree crown extraction using airborne LiDAR data. *ISPRS Journal of Photogrammetry and Remote Sensing*, 104: 30–43.

Sun, G., Ranson, K.J., Guo, Z., Zhang, Z., Montesano, P., and Kimes, D., 2011. Forest biomass mapping from lidar and radar synergies. *Remote Sensing of Environment*, 115: 2906–2916.

Tang, S., Dong, P., and Buckles, B., 2013. Three-dimensional surface reconstruction of tree canopy from LiDAR point clouds using a region-based level set method. *International Journal of Remote Sensing*, 34: 1373–1385.

Van Leeuwen, M., Coops, N.C., and Wulder, M.A., 2010. Canopy surface reconstruction from a LiDAR point cloud using Hough transform. *Remote Sensing Letters*, 1: 125–132.

Vastaranta, M., Holopainen, M., Yu, X., Haapanen, R., Melkas, T., Hyyppä, J., and Hyyppä, H., 2011. Individual tree detection and area-based approach in retrieval of forest inventory characteristics from low-pulse airborne laser scanning data. *Photogrammetric Journal of Finland*, 22: 1–13.

Vega, C., Hamrouni, A., El Mokhtari, S., Morel, J., Bock, J., Renaud, J.-P., Bouvier, M., and Durrieu, S., 2014. PTrees: A point-based approach to forest tree extraction from lidar data. *International Journal of Applied Earth Observation and Geoinformation*, 33: 98–108.

Wang, C., and Glenn, N.F., 2008. A linear regression method for tree canopy height estimation using airborne lidar data. *Canadian Journal of Remote Sensing*, 34: S217–S227.

Wang, C., and Glenn, N.F., 2009. Integrating LiDAR intensity and elevation data for terrain characterization in a forested area. *IEEE Geoscience and Remote Sensing Letters*, 6: 463–466.

Weinacker, H., Koch, B., Heyder, U., and Weinacker, R., 2004. Development of filtering, segmentation and modeling modules for LIDAR and multispectral data as a fundament of an automatic forest inventory system. *International Archives of Photogrammetry, Remote Sensing and Spatial Information Sciences*, XXXVI (8/W2): 50–55.

Wulder, M., Niemann, K.O., and Goodenough, D.G., 2000. Local maximum filtering for the extraction of tree locations and basal area from high spatial resolution imagery. *Remote Sensing of Environment*, 73: 103–114.

Wulder, M.A., White, J.C., Nelson, R.F., Næsset, E., Ole Øroa, H., Coops, C.N., Hiller, T., Bater, C.W., and Gobakken, T., 2012. LiDAR sampling for large-area forest characterization: A review. *Remote Sensing of Environment*, 121: 196–209.

Yu, X., Hyyppä, J., Holopainen, M., and Vastaranta, M., 2010. Comparison of area-based and individual tree-based methods for predicting plot-level forest attributes. *Remote Sensing*, 2: 1481–1495.

Zhang, K., Simard, M., Ross, M., Rivera-Monroy, V.H., Houle, P., Ruiz, P., Twilley, R.R., and Whelan, K.R.T., 2008. Airborne laser scanning quantification of disturbances from hurricanes and lightning strikes to mangrove forests in Everglades National Park, USA. *Sensors*, 8: 2262–2292.

Zhao, K., Popescu, S., and Nelson, R., 2009. LiDAR remote sensing of forest biomass: A scale-invariant estimation approach using airborne lasers. *Remote Sensing of Environment*, 113: 182–196.

Zhao, K., Popescu, S., Meng, X., Pang, Y., and Agca, M., 2011. Characterizing forest canopy structure with LiDAR composite metrics and machine learning. *Remote Sensing of Environment*, 115: 1978–1996.

Zolkos, S.G., Goetz, S.J., and Dubayah, R., 2013. A meta-analysis of terrestrial aboveground biomass estimation using lidar remote sensing. *Remote Sensing of Environment*, 128: 289–298.

5 LiDAR for Urban Applications

5.1 INTRODUCTION

LiDAR has been widely used in many studies in urban environments. In the review by Yan et al. (2015a) on LiDAR for urban land cover classification, five applications were discussed: (1) urban morphology and green analysis, (2) urban flood risk modeling, (3) mapping power transmission lines, (4) modeling GPS/airport signal obstruction, and (5) solar radiation assessment. Other applications of LiDAR in urban environments include building extraction and modeling (Haala and Brenner 1999, Maas and Vosselman 1999, Stilla and Jurkiewicz 1999, Brenner 2005, Madhavan et al. 2006, Alexander et al. 2009, Jang and Jung 2009, Pu and Vosselman 2009, Wang 2013, Xiao et al. 2014), road extraction (Boyko and Funkhouser 2011, Kumar et al. 2013, Landa and Prochazka 2014, Li et al. 2015), impervious surface mapping (Hodgson et al. 2003, Cui 2014), population estimation (Dong et al. 2010, Qiu et al. 2010, Xie et al. 2015, Zhao et al. 2017), change detection (Singh et al. 2012, Teo and Shih 2012, Zavodny 2012, Hebel et al. 2013), post-earthquake assessment of building damage (Li et al. 2008, Hussain et al. 2011, Dong and Guo 2012, Dong and Shan 2013), and post-earthquake assessment of road blockage (Liu et al. 2014). This chapter presents six major topics of LiDAR applications in urban environments: (1) road extraction, (2) building extraction and 3D reconstruction, (3) population estimation, (4) change detection, (5) assessment of post-disaster building damage, and (6) assessment of post-disaster road blockage. At the end of the chapter, two step-by-step projects in ArcGIS are presented to showcase LiDAR-based powerline corridor analysis and small-area population estimation using building count, building area, and building volume in Denton, TX, USA.

5.2 ROAD EXTRACTION

For updating GIS databases and other applications, road extraction from digital images has received considerable attention in the past decades. Increasingly, LiDAR data has been used for road extraction (e.g., Boyko and Funkhouser 2011, Kumar et al. 2013, Landa and Prochazka 2014, Li et al. 2015, Ferraz et al. 2016). Fully automated road extraction in urban areas can be difficult due to the complexity of urban features, while manual digitizing of roads from images can be time-consuming. In many cases, a semi-automated approach to road extraction can be implemented to improve the efficiency, accuracy, and cost-effectiveness of data development activities.

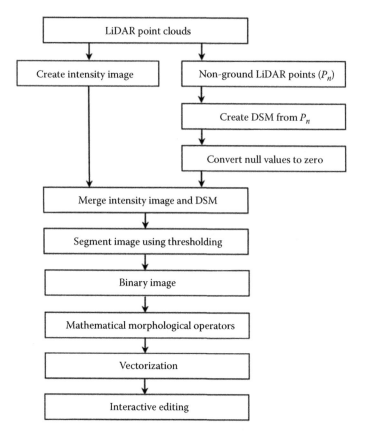

FIGURE 5.1   Flowchart for road extraction using LiDAR data.

Ground features such as water bodies and asphalt pavement usually have very low LiDAR intensity values, while some building roofs may also have very low intensity values. Therefore, integration of LiDAR intensity data and digital surface models (DSM) or digital height models (DHM) can be used for road extraction. Figure 5.1 shows a flowchart for road extraction using airborne LiDAR point clouds. First, an intensity image is created from LiDAR point clouds using spatial interpolation; then a DSM is created from non-ground LiDAR points, and null values in the DSM are converted to zeros. After merging the intensity image and non-ground DSM, the merged image can be segmented using a threshold. The resulting binary image can be further processed using mathematical morphological operators (Serra 1983, Dong 1997). After vectorization, the extracted roads can be corrected through interactive editing. Figure 5.2 is a LiDAR intensity image created from 2009 LiDAR data in the city of Denton, TX, USA; Figure 5.3 shows the non-ground DSM; Figure 5.4 is the merged image created by adding the intensity image and the non-ground DSM; and Figure 5.5 is the binary image showing the roads and some other ground features.

FIGURE 5.2   LiDAR intensity image for an area in Denton, TX, USA.

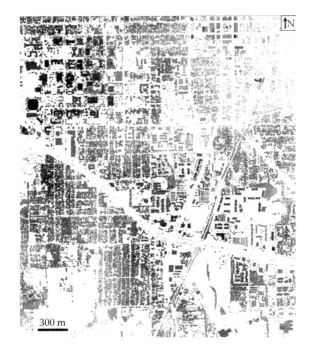

FIGURE 5.3   Non-ground DSM derived from LiDAR data.

FIGURE 5.4   A merged image is created by adding LiDAR intensity image and non-ground DSM.

5.3   BUILDING EXTRACTION AND 3D RECONSTRUCTION

With the increasing availability of high-resolution satellite images and LiDAR data in the last two decades, extracting buildings from remotely sensed data has become a significant research field. Many algorithms for building extraction have been proposed, including mathematical morphology (Weidner and Förstner 1995, Pesaresi and Benediktsson 2001, Mongus et al. 2013), DSM segmentation (Baltsavias et al. 1995, Sithole and Vosselman 2004, Tovari and Pfeifer 2005, Awrangjeb and Fraser 2014), active contours or snakes (Nixon and Aguado 2002, Oriot 2003, Ahmadi et al. 2010, Kabolizade et al. 2010, Yan et al. 2015b), the Dempster–Shafer method (Rottensteiner et al. 2005, Trinder et al. 2010), neural networks (Barsi 2004, Bellman and Shortis 2004, Liu et al. 2013), knowledge-based systems (Baltsavias 2004, Mayer 2008, Susaki 2013), and the multi-scale method (Vu et al. 2009, Zhang et al. 2012). A review of most of the methods can be found in the work of Ioannidis et al. (2009). With the increasing demand for three-dimensional (3D) city models and availability of LiDAR data, 3D building reconstruction has received extensive attention, and many methods for building reconstruction have been proposed (Gruen 1998, Haala and Brenner 1999, Maas and Vosselman 1999, Stilla and Jurkiewicz 1999, Stilla et al. 2003, Suveg and Vosselman 2004, Brenner 2005, Madhavan et al. 2006, Sugihara and Hayashi 2008, Alexander et al. 2009, Jang and Jung 2009, Pu and Vosselman 2009, Orthuber and Avbelj 2015).

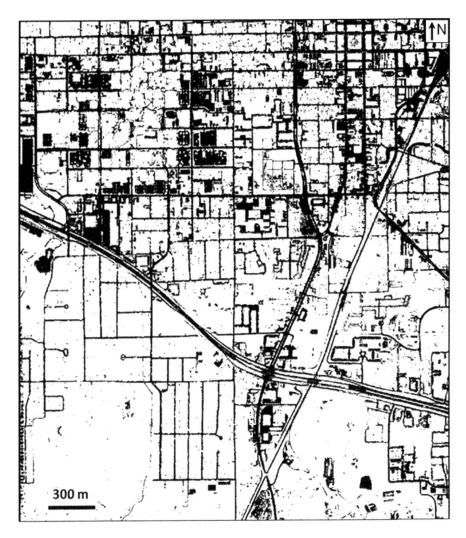

FIGURE 5.5 Binary image for extracted roads and other surface features. Further processing is needed to obtain topologically correct road networks.

Xiao et al. (2014) described a Gaussian mixture model for building roof segmentation. The Gaussian distribution is widely used in the natural sciences as a simple model for describing complex phenomena with a variety of probability distributions. The Gaussian distribution defined over a d-dimensional vector x of continuous variables is shown by:

$$N(x\,|\,\mu, \Sigma) = \frac{1}{(2\pi)^{d/2}} \frac{1}{|\Sigma|^{1/2}} \exp\left\{-\frac{1}{2}(x-\mu)^T \Sigma^{-1}(x-\mu)\right\} \quad (5.1)$$

where the d-dimensional vector μ is the mean and the $d \times d$ matrix Σ is the covariance. If x denotes the 3D coordinates of a LiDAR point, then the 3D Gaussian model can be used to describe the distribution of LiDAR points within a planar patch. Due to the limited accuracy of LiDAR data, LiDAR point clouds of planar roofs do not exactly lie on a mathematical plane but scatter within a thin plate near the roof, and the deviations of the points from the plane conform to a Gaussian distribution centered at zero. Therefore, a single building with a multi-plane roof is represented by a 3D Gaussian mixture model, which is a linear superposition of Gaussian distributions in the following form:

$$p(x) = \sum_{k=1}^{K} \pi_k N(x \mid \mu_k, \Sigma_k) \qquad (5.2)$$

where x is observed data vector, K is the number of Gaussian components in the Gaussian mixture model, $N(\mu_k, \Sigma_k)$ is the kth Gaussian distribution of the Gaussian mixture model, μ_k and Σ_k are the mean and covariance of the kth component respectively, and π_k is the mixture coefficients. The mixture coefficients π_k satisfy $\sum_{k}^{K} \pi_k = 1$, $0 \le \pi_k \le 1$.

Figure 5.6 shows comparisons of detected roof planes using the Gaussian mixture model (Xiao et al. 2014) and the popular RANSAC (random sample consensus) method proposed by Fischler and Bolles (1981). As can be seen, for building 1 (Figure 5.6A), the Gaussian mixture model detects two roof-plane intersections (Figure 5.6B), whereas RANSAC only detects two roof planes, and fails to detect roof-plane intersections (Figure 5.6C). For building 2 (Figure 5.6D), the Gaussian

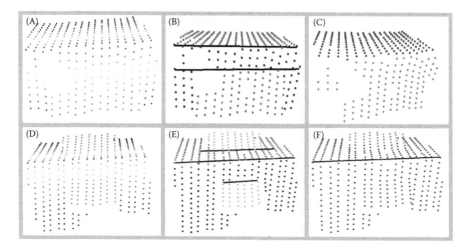

FIGURE 5.6 Comparison of roof planes extracted from Gaussian mixture model and RANSAC model. (A) Raw LiDAR points for building 1; (B) roof planes extracted from (A) by Gaussian mixture model; (C) roof planes extract from (A) by RANSAC method; (D) raw LiDAR points for building 2; (E) roof planes extracted from (D) by Gaussian mixture model; and (F) roof planes extract from (D) by RANSAC method.

mixture model successfully detects three roof-plane intersections (Figure 5.6E), whereas the RANSAC method only detects one roof-plane intersection (Figure 5.6F).

5.4 POPULATION ESTIMATION

The traditional census method for obtaining population information is often time-consuming, labor-intensive, and expensive. Therefore, many national census agencies and international organizations use four methods to update census data and estimate population size (Smith and Mandell 1984): (1) Component II—using vital statistics such as birth and death data to measure the natural increase from the last census; (2) Ratio-correlation—using regression methods to relate changes in population to changes in indicators of population change, such as school enrolment, the number of voters, the number of passenger car registrations, and the number of occupied housing units; (3) Administrative record—using births, deaths, school enrolment, social insurance, building permits, driver licenses, voter registration, and tax returns to estimate population size; and (4) Cohort component method—tracing people born in a given year through their lives. Numerous studies have been carried out for population estimation using demographic characteristics (Verma et al. 1984, Platek et al. 1987, Bracken 1991, Wolter and Causey 1991, Cai 2007, Jarosz 2008).

Population estimation using remote sensing and GIS belongs to the ratio-correlation method, and can be classified into two broad categories: areal interpolation and statistical modeling (Figure 5.7). Since the 1970s, remotely sensed data from various platforms have been used for population estimation, including low-resolution images (Welch and Zupko 1980, Sutton et al. 1997, 2001), medium-resolution images (Harvey 2002, 2003, Lo 2003, Qiu et al. 2003, Wu and Murray 2003, Wu 2004, Li and Weng 2005, Lu et al. 2006), and high-resolution aerial photographs (Lo and Welch 1977, Lo 1986 a,b).

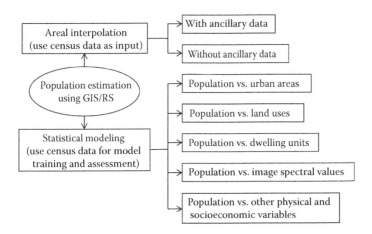

FIGURE 5.7 Methods of population estimation using GIS and remote sensing. (From Dong, P. et al., *Int. J. Remote Sens.*, 31, 5571–5586, 2010.)

Dong et al. (2010) presented a method for small-area population estimation using LiDAR, Landsat TM, and parcel data. The basic idea is to build ordinary least squares (OLS) models and geographically weighted regression (GWR) models (Lo 2008) based on sampling census blocks using (1) population vs. building count, (2) population vs. building area, and (3) population vs. building volume. Figure 5.8 shows a flowchart of population estimation using Landsat TM, LiDAR, parcels, and census data in an area in the city of Denton, TX, USA.

Using census population counts from 91 random census blocks as the dependent variable, and LiDAR-derived building count, building area, and building volume as independent variables, linear regression models and GWR were created for population estimation in the study by Dong et al. (2010). The results suggest that population count is strongly correlated with residential building count, area, and volume derived

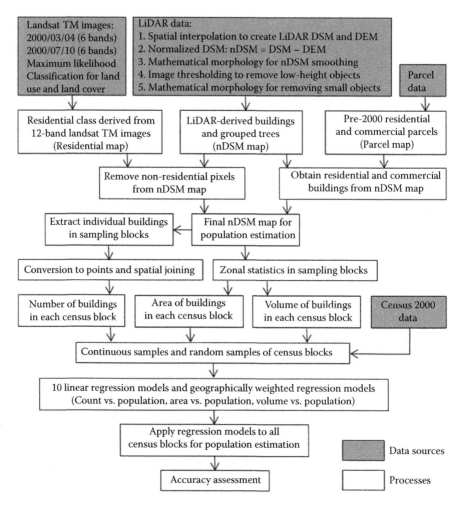

FIGURE 5.8 Flowchart of population estimation using GIS and remotely sensed data. (From Dong, P. et al., *Int. J. Remote Sens.*, 31, 5571–5586, 2010.)

from the final DHM. Although other studies show that the traditional housing unit method offers a number of advantages over other population estimation methods (Smith and Lewis 1980, 1983, Smith and Cody 1994), there is no obvious pattern to show that building counts outperformed the other two independent variables (building area and building volume) in this study. In addition, as random samples can better represent data distribution in the study area than clustered samples, models derived from random samples seem to generate more accurate results compared with those derived from continuous samples. When the spatial heterogeneity is taken into account, the GWR models provide more accurate estimates than the linear regression models.

A limitation in the study by Dong et al. (2010) is that the original LiDAR data used was resampled to a point spacing of 3–5 m, which affects accurate representation of buildings. More accurate results could be obtained if LiDAR data with higher point density was available. To help readers practice estimating population using LiDAR-derived building count, building area, and building volume, Project 5.1 demonstrates a complete process in ArcGIS for population estimation using a 2009 LiDAR dataset with a point density of approximately 1 point/m^2 acquired in the city of Denton, TX, USA, and a 2010 census block shapefile with census data.

5.5   CHANGE DETECTION

Change detection using pixel-based analysis of remotely sensed data has been well documented (e.g., Singh 1989, Mouat et al. 1993, Deer 1995, Coppin et al. 2004, Lu et al. 2004). The pixel-based change detection methods have been less successful when applied to high-resolution or very-high-resolution (VHR) images due to issues with georeferencing accuracy, reflectance variability, and data acquisition geometry, among other factors. The increasing computational capabilities and availability of VHR images in the last two decades have prompted the emergence of object-based image analysis (OBIA) from the traditional pixel-based image analysis (e.g., Longley 2002, Aplin and Smith 2008, Blaschke 2010, Stow 2010, Addink et al. 2012). A review of change detection from remotely sensed images from pixel-based to object-based approaches can be found in the work of Hussain et al. (2013).

Various studies have demonstrated that change detection is a complex process, and no single approach is applicable to all application scenarios. For example, OBIA generally requires image segmentation based on spectral and spatial information of the image, and the results are affected by segmentation algorithms (Hay et al. 2005, Kim et al. 2008, Lein 2012). Integration of GIS and remote sensing can be an effective approach for improving change detection because spatial and attribute information of objects are conveniently combined with spectral information (Li 2010). For example, Wu et al. (2014) used the energy spectrum of the windowed Fourier transform of each land-use parcel for parcel-based change detection.

With the increasing use of LiDAR data in urban studies, 3D change detection of horizontal and vertical urban sprawl has become an important topic (Teo and Shih 2012, Singh et al. 2012). Dong et al. (2018) proposed a method for parcel-based building change detection to support applications such as urban planning and land management. The method uses digital terrain models (DTM), DSM,

DHM (DHM = DSM − DTM), and differenced digital surface models (dDSM). After a series of processing steps in GIS, the proposed method produces an output building map to show four types of buildings—Type I (new buildings that are built after removing medium/high vegetation), Type II (new buildings that are built on bare earth or low vegetation, or on top of existing buildings), Type III (demolished or damaged buildings), and Type IV (existing buildings that have little or no changes).

Figure 5.9 shows LiDAR point clouds collected in March 2009 and April 2013 in a 1 km × 1 km area in the city of Surrey, British Columbia, Canada. The 2009 dataset has a point density of about 2 points/m², with four returns and five classes, including 2 (ground), 7 (noise), 9 (water), 12 (overlap), and 21 (reserved). The 2013 dataset has a point density of 25–30 points/m², with five returns and eight classes, including 2 (ground), 3 (low vegetation, less than 0.7 m), 4 (medium vegetation, 0.7 to 2 m), 5 (high vegetation, above 2 m), 6 (building), 7 (noise), 9 (water), and 11 (reserved). Figure 5.10 shows that trees in a point cloud profile derived from 2009 LiDAR data have become buildings in a point cloud profile derived from 2013 LiDAR data of the same area. Since LiDAR points for vegetation and building classes are readily available in the 2013 data, but not in the 2009 data, changes in vegetation and buildings from 2009 to 2013 cannot be obtained by direct comparison, and a change detection method needs to be developed.

The flowchart for parcel-based building change detection proposed by Dong et al. (2018) is shown in Figure 5.11, and the major steps for the flowchart are explained below.

1. To determine a proper slope threshold separating rooftops and medium/ high vegetation, both zonal statistics and focal statistics (neighborhood statistics) are used. Figure 5.12A shows the histogram of zonal mean rooftop slopes obtained from 1112 building footprint polygons from the 2013

(A)

(B)

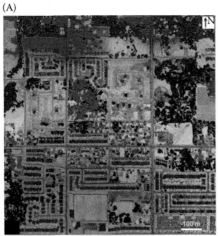

FIGURE 5.9   LiDAR point clouds in LAS format for 2009 (A), and 2013 (B).

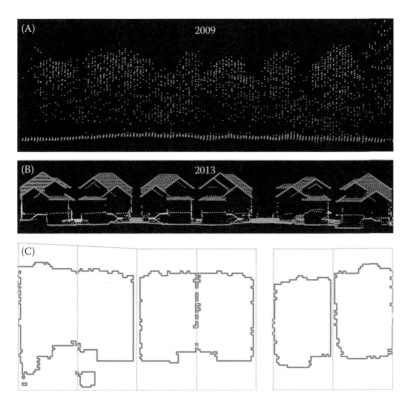

FIGURE 5.10 Samples of change detection using multi-temporal LiDAR data in Surrey, British Columbia, Canada. (A) Point cloud profile of trees derived from 2009 LiDAR data; (B) Point cloud profile of buildings derived from 2013 LiDAR data; (C) Building footprints (red) derived from 2013 LiDAR data for each parcel (green).

building slope raster data. Figure 5.12B is the histogram of focal mean surface slope obtained from 21 × 21 (10.5 m × 10.5 m) windows in the 2009 LiDAR DHM at 734 random locations over medium/high vegetation areas. From Figure 5.12, it can be concluded that a 2013 building footprint polygon highly likely contained medium/high vegetation in 2009 if the mean slope of the 2009 DHM in the same polygon is greater than the maximum zonal mean slope of 2013 rooftops. This is how Type I building changes are identified in Figure 5.11. To accommodate possible zonal mean rooftop slopes in the 2009 slope raster that are greater than the maximum zonal mean rooftop slopes in the 2013 slope raster S_{max} (51.11° in this case), the slope threshold S_t is set as $S_{max} + 2$. In other words, 53.11° can be used as the threshold (S_t in Figure 5.11) for zonal mean slope to separate rooftop and medium/high vegetation. If $S_i > S_t$ (see Figure 5.11), the objects derived from the 2009 LiDAR data are medium/high vegetation, representing Type I building change—new buildings are built after removing medium/high vegetation.

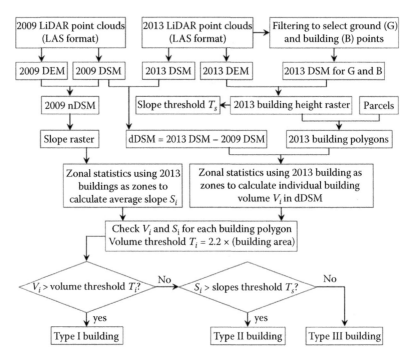

FIGURE 5.11 Flowchart for parcel-based building change detection using 2009 and 2013 LiDAR data. Shaded boxes represent input data for the flowchart. S_i is the zonal mean slope for building polygon i, and V_{ij} is the volumetric change from 2009 to 2013 for building polygon i. MMO standards for mathematical morphological operations. (From Dong, P. et al. *Surv. Land Inf. Sci.*, 2018.)

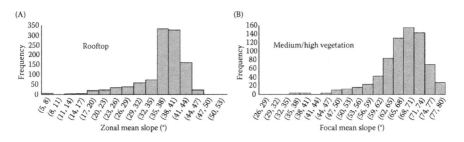

FIGURE 5.12 Histograms of zonal mean rooftop slopes from 2013 data (A), and focal mean surface slopes for medium/high vegetation in 2009 data (B). (From Dong, P. et al. *Surv. Land Inf. Sci.*, 2018.)

2. Calculate the zonal mean slope S_i from the slope raster created in Step 2, using the 2013 building footprint polygons as zones. Note S_i can be for any object, such as buildings, trees, or bare earth in the 2009 DHM.

3. Denote V_{ij} as the volumetric change of ground objects (buildings, trees, ground surfaces, etc.) from 2009 to 2013 and then calculate V_{ij} from dDSM using the zonal sum of dDSM cells in each of the 2013 building footprint

polygons. Note V_{ij} can be for any ground object (buildings, trees, ground surfaces, etc.), and it can be positive, zero, or negative.

4. Check S_i and V_{ij} for each of the 2013 building footprint polygons. A volume threshold V_t is used in this process, and $V_t = 2.2$ * (building footprint area), where 2.2 is the minimum height (in meters) of a residential building based on a home construction and safety standard used in Texas (Dong et al. 2010), because a similar stand was not found for British Columbia.

5. The decision process for identifying the four types of building change is illustrated in Figure 5.13. It should be noted that the threshold for volumetric changes, especially between Type III and Type IV, is subjectively selected for simplicity's sake. For example, the boundary between "damaged" and "minor changes" can be difficult to define objectively. Similarly, there is no clear threshold between "demolished" and "damaged," and the two terms can be used for building changes in different application scenarios, such as land use change detection or disaster damage assessment.

Visual comparison of the 2009 and 2013 LiDAR-derived DHMs indicate that the results for building change types in Figure 5.14 are correct in most cases. To better understand the performance of the method, all 1112 buildings extracted from the 2013 LiDAR data were visually compared with the 2009 DHM using three quantitative measures: completeness, correctness, and quality (Heipke et al. 1997).

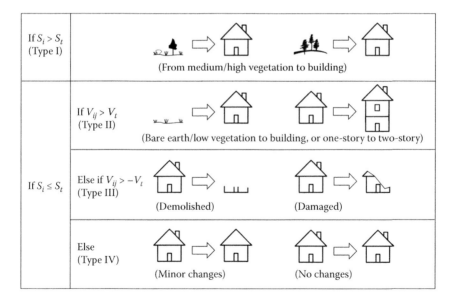

FIGURE 5.13 Identifying four types of building change using zonal mean rooftop slope (S_i), slope threshold (S_t), zonal volumetric change (V_{ij}), and volume threshold (V_t). (From Dong, P. et al. *Surv. Land Inf. Sci.*, 2018.)

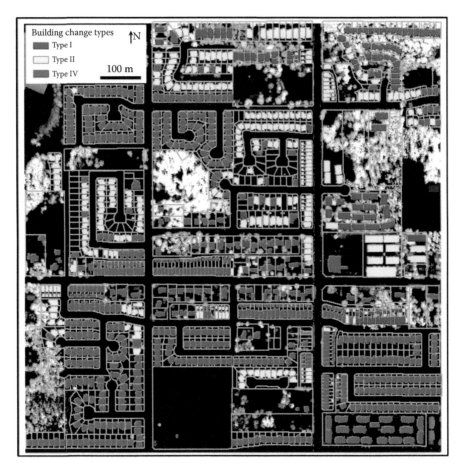

FIGURE 5.14 Map of building change types. Type III is not detected in the study area. The 2009 DHM is used as the backdrop. (From Dong, P. et al. *Surv. Land Inf. Sci.*, 2018.)

$$Completeness = \frac{TP}{TP + FN} \qquad (5.3)$$

$$Correctness = \frac{TP}{TP + FP} \qquad (5.4)$$

$$Quality = \frac{TP}{TP + FN + FP} \qquad (5.5)$$

where TP is true positive, FN is false negative, and FP is false positive. Table 5.1 shows that the results for the three building change types are very accurate. Further analyses of the results suggest that errors are mainly caused by buildings under construction, buildings under tree canopy, and relatively low density of the 2009 LiDAR points (Dong et al. 2018).

TABLE 5.1

Completeness, Correctness, and Quality Measures for Three Types of Building Change

Change Type	Completeness	Correctness	Quality
Type I	0.96	0.93	0.90
Type II	0.97	0.91	0.89
Type IV	0.97	0.98	0.95

Source: Dong, P. et al. *Surv. Land Inf. Sci.*, 2018.

5.6 ASSESSMENT OF POST-DISASTER BUILDING DAMAGE

Rapid assessment of damages to buildings and infrastructure caused by natural disasters such as earthquakes is essential for disaster response and recovery operations. Although real-time support for ground search and rescue is still a challenging task for current remote sensing systems, data collected by remote sensing platforms has been widely used in disaster damage assessment. Medium- and high-resolution optical images such as Système Probatoire d'Observation de la Terre (SPOT) high resolution visible (HRV), QuickBird, and IKONOS images have been used for post-earthquake building damage assessment with limited success (Aoki et al. 1998, Huyck et al. 2002, Matsuoka and Yamazaki 2002, Yusuf et al. 2002, Adams 2004, Saito and Spence 2004, Adams et al. 2005, 2006, Kaya et al. 2005). Major limitations of optical images in damage assessment include difficulties in data collection in bad weather and time-consuming processes in data analysis. The all-weather capability of radar systems provides an important option for damage assessment in bad weather (Guo et al. 2009, Wang et al. 2009). However, radar layover and shadow effects can hamper image interpretation, particularly in mountain environments. Interferometric synthetic aperture radar (InSAR) could also be an alternative data source for damage assessment because of its ability to provide elevation data, but the data quality and the elevation accuracy derived from InSAR are lower than LiDAR data (Stilla and Jurkiewicz 1999, Stilla et al. 2003).

To evaluate potential applications of LiDAR point clouds for post-earthquake building damage assessment, Dong and Guo (2012) simulated soft-story collapse and other types of major damage to buildings with flat, pent, gable, and hips roofs (Figure 5.15). Triangulated irregular networks (TINs) of simulated LiDAR points were created through Delaunay triangulation. For a triangle with vertices p_1, p_2, and p_3, any point p in the triangle can then be specified by a weighted sum of these three vertices; that is:

$$p = t_1 p_1 + t_2 p_2 + t_3 p_3 \qquad (5.6)$$

where $t_1 + t_2 + t_3 = 1$, and t_1, t_2, and t_3 are called barycentric coordinates (Coxeter 1969). If the 3D Cartesian coordinates of p_1, p_2, p_3, and p are (x_1, y_1, z_1), (x_2, y_2, z_2),

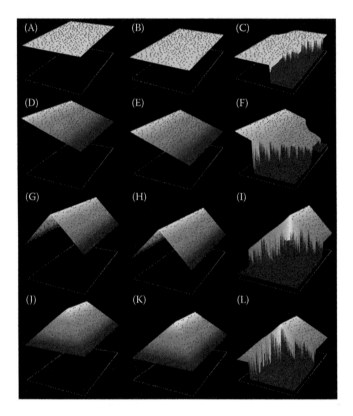

FIGURE 5.15 Simulated 3D building models with flat (first row), pent (second row), gable (third row) and hips roofs (fourth row). First column: three-story models; second column: two-story models (for soft-story collapse of three-story models); third column: damaged models. Models for total collapse are not shown here. (From Dong, P. and Guo, H., *Int. J. Remote Sens.*, 33, 81–100, 2012.)

(x_3, y_3, z_3), and (x, y, z) respectively, x, y, z can be calculated using the following equations:

$$x = t_1 x_1 + t_2 x_2 + (1 - t_1 - t_2) x_3 \qquad (5.7)$$

$$y = t_1 y_1 + t_2 y_2 + (1 - t_1 - t_2) y_3 \qquad (5.8)$$

$$z = t_1 z_1 + t_2 z_2 + (1 - t_1 - t_2) z_3 \qquad (5.9)$$

In computer implementation, t_1 and t_2 are random numbers between 0 and 1. If $t_1 + t_2 > 1$, then t_1 is replaced with $1 - t_1$, and t_2 replaced with $1 - t_2$. This is to ensure that barycentric coordinates will be uniformly distributed in the triangle instead of creating clusters. By generating random points on the buildings walls and surfaces (Figure 5.16), 3D shape signatures of the building models can be compared to detect damaged buildings (Figure 5.17). The method is also applied to real LiDAR data

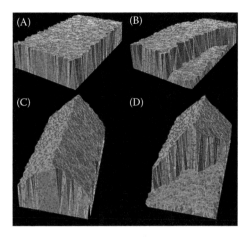

FIGURE 5.16 New random points generated using the TIN model and barycentric coordinates. (A) Original flat roof model, (B) damaged flat roof model, (C) original gable roof model, and (D) damaged gable roof model. (From Dong, P. and Guo, H., *Int. J. Remote Sens.*, 33, 81–100, 2012.)

for several buildings in Harris County, TX, USA (Figure 5.18). In Figure 5.18A, the upper part shows a TIN model built from LiDAR data for six buildings with a point density of about 0.8 point/m^2, and the lower part is an air photo of the six buildings. The 3D shape signatures derived from the LiDAR data for buildings 1, 2, 4 and 5 are shown in Figure 5.18B, and the correlation coefficients between the 3D shape signatures are listed in Table 5.2. The results from 3D shape signature analyses are in accord with those from visual interpretation of the air photo: buildings 1 and 5 are the same model, while buildings 2 and 4 belong to a different model. This example shows that major changes in 3D building shapes can be detected by 3D shape signatures obtained from LiDAR data. Following the same logic, if pre-earthquake 3D shape signatures of a building are known, it is possible to detect severe damage or collapse of the building by comparing its pre-earthquake 3D shape signatures with post-earthquake 3D shape signatures derived from LiDAR data.

Based on literature review and new results in building damage assessment using LiDAR data, Dong and Guo (2012) proposed a framework with four major components for automated assessment of post-earthquake building damage using LiDAR data and GIS (Figure 5.19). Compared with the tiered reconnaissance system proposed by Adams et al. (2004), the framework focuses on per-building (Tier 3) assessment. Assessment at regional (Tier 1) and neighborhood (Tier 2) levels can be incorporated later to form a broader framework. Figure 5.20 is a flowchart to help understand the framework.

5.7 ASSESSMENT OF POST-DISASTER ROAD BLOCKAGE

In addition to the building damage assessment previously described, infrastructure damage assessment is also an important component of post-disaster damage assessment. With the development of Volunteered Geographic Information (VGI) (Goodchild 2007, Elwood 2008, Flanagin and Metzger 2008) and NeoGeography

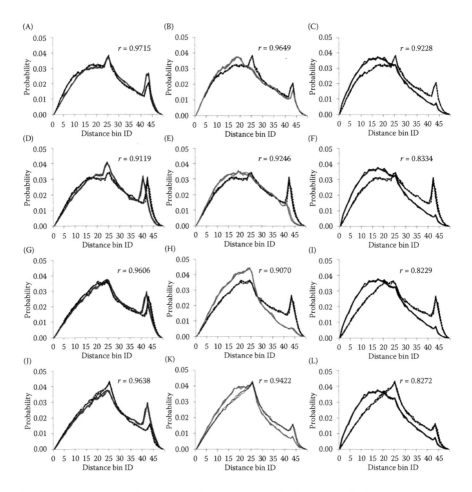

FIGURE 5.17 Comparisons of 3D shape signatures obtained from the building models in Figure 5.11 using the TIN model and barycentric coordinate system. First row: flat roofs; second row: pent roofs; third row: gable roofs; and fourth row: hip roofs. First column: three-story models vs. two-story models; second column: two-story models vs. damaged models; third column: two-story models vs. collapsed models. r is the correlation coefficient between the 3D shape signature curves of two models. (From Dong, P. and Guo, H., *Int. J. Remote Sens.*, 33, 81–100, 2012.)

(Goodchild 2009), grassroots citizens can make contributions to the global geospatial data infrastructure, as demonstrated in the OpenStreetMap data collection activities after the Haiti earthquake in 2010. However, road centerlines collected by internet users worldwide may contain positional errors and connection errors in the road network.

To detect road connection errors and post-disaster road blockage points, Liu et al. (2014) developed a web-based application that allows users to create road centerlines based on high-resolution images, and detect road blockage using LiDAR data. Figure 5.21 is a sample application for Port-au-Prince, Haiti. Using a transparent

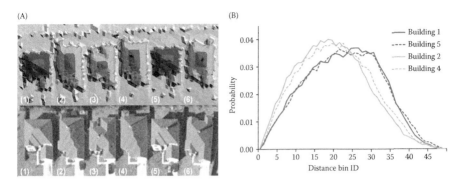

FIGURE 5.18   3D shape signature analysis using LiDAR data for selected buildings in Harris County, TX, USA. (A) TIN model built from LiDAR (top), and air photo (bottom); the numbers in brackets are the building numbers and (B) 3D shape signatures derived from LiDAR data of selected buildings. (From Dong, P. and Guo, H., *Int. J. Remote Sens.*, 33, 81–100, 2012.)

TABLE 5.2
Correlation Coefficients between 3D Shape Signatures of Selected Buildings

	Building 1	Building 2	Building 4	Building 5
Building 1	1	0.8702	0.9271	**0.9962**
Building 2	0.8702	1	**0.9902**	0.8726
Building 4	0.9271	**0.9902**	1	0.9308
Building 5	**0.9962**	0.8726	0.9308	1

Source: Dong, P. and Guo, H., *Int. J. Remote Sens.*, 33, 81–100, 2012.
Note: Bold numbers show high correlation coefficients between the building models of the same type.

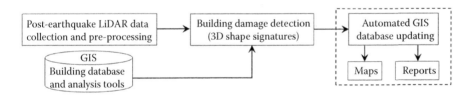

FIGURE 5.19   A framework for automated assessment of post-earthquake building damage using LiDAR data and GIS. (From Dong, P. and Guo, H., *Int. J. Remote Sens.*, 33, 81–100, 2012.)

display method, Figure 5.21 shows a high-resolution satellite image and a background digital surface model derived from post-earthquake LiDAR data. Road blockage points (red points in Figure 5.21) for selected road segments can be detected. In addition, errors in the OpenStreetMap road centerlines can be detected. For example, the red point in the upper-left corner of Figure 5.22 is in fact a wall separating the highway and the residential community, and is detected as a blockage point. However, other detected blockage points are not validated due to the difficulty in field data collection.

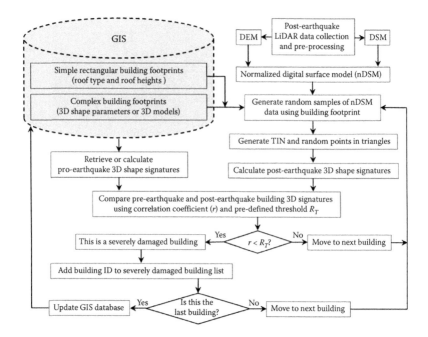

FIGURE 5.20 Flowchart for automated assessment of post-earthquake building damage. (From Dong, P. and Guo, H., *Int. J. Remote Sens.*, 33, 81–100, 2012.)

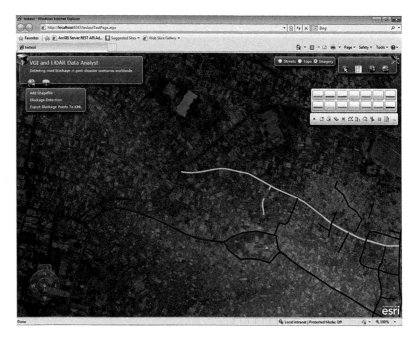

FIGURE 5.21 Web-based application for post-disaster road blockage detection using high-resolution images and LiDAR data.

FIGURE 5.22   Road blockage and connection errors (red points) detected in Port-au-Prince, Haiti using web-based application. The background is a LiDAR-derived digital surface model.

PROJECT 5.1:   LOCATING 3D LiDAR POINTS NEAR A POWERLINE IN THE CITY OF DENTON, TX, USA

1. Introduction

Esri's 3DGISTeam created 3D Sample Tools for ArcGIS Desktop 10.2–10.5 in February 2015. The 3D Sample Tools have over 50 geoprocessing tools and stand-alone utilities for data conversion, feature analysis, TIN, vegetation analysis, LiDAR analysis, LiDAR classification, and LiDAR management (Figure 5.23).

2. Data

The following files are available at http://geography.unt.edu/~pdong/ LiDAR/Chapter5/Project5.1/: (1): An LAS file "Denton2011.las" for LiDAR point clouds covering an area of 2.97 km × 3.15 km in Denton, TX, USA; and (2) A 3D shapefile "PowerLine3D.shp" (and seven other files with different extensions) simulating a powerline in 3D. The location of the simulated powerline is along West Hickory Street and between Avenue C and Avenue F in Denton, TX, USA. These files should be downloaded to your local folder (right click each file to download the files).

3. Project Steps

   1. Open an empty Word document so that you can copy any results from the following steps to the document. To copy the whole screen to your Word document, press the PrtSc (print screen) key on your keyboard, then open your Word document and click the "Paste" button or press

FIGURE 5.23   3D Sample Tools in ArcToolbox.

Ctrl+V to paste the content into your document. To copy an active window to your Word document, press Alt+PrtSc, then paste the content into your document.

2. Open ArcMap and check if the 3D Sample Tools are listed in the ArcToolbox. If the 3D Sample Tools are already installed in your ArcToolbox, go to Step (3). Otherwise, you can download the tools from http://www.arcgis.com/home/item.html?id=fe221371b77940749f f96e90f2de3d10 (if this link does not exist anymore, you can download "Install_3D_Samples.zip" from http://geography.unt.edu/~pdong/ LiDAR/Chapter5/Project5.1/), unzip the downloaded zip file, and run "Install 3D Sample" to install the 3D Sample Tools in a folder. Right click ArcToolbox and select "Add Toolbox…", then select "3D Sample Tools.tbx" from the installation folder, and click "Open" (Figure 5.24). The 3D Sample Tools should appear in the ArcToolbox.

3. Open ArcMap, select Customize → Toolbars and select "LAS Dataset," then select Customize → Extensions and turn on the 3D Analyst Extension.

4. Open ArcToolbox → Data Management Tools → LAS Dataset → Create LAS Dataset. Use Denton2011.las as input and Denton.lasd as output to create a LiDAR dataset. The LAS dataset is added to ArcMap automatically.

5. Add PowerLine3D to ArcMap, right click "PowerLine3D" and select "Zoom To Layer," and then save your ArcMap project as Project5.1.mxd (Figure 5.25).

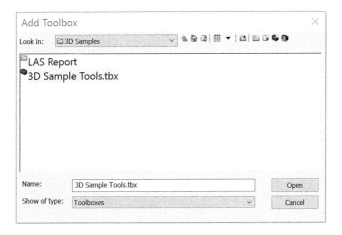

FIGURE 5.24   Adding 3D Sample Tools to ArcToolbox.

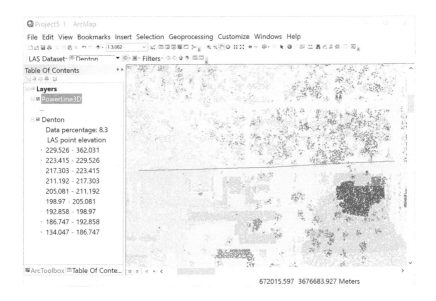

FIGURE 5.25   LiDAR dataset and a simulated powerline in Denton, TX, USA.

6. Open ArcToolbox → 3D Sample Tools → Lidar Analysis → Locate LAS Points By Proximity. Use Denton.lasd as input LAS dataset, PowerLine3D as input 3D features, 5 as 3D distance, Proximity5.shp as output feature class, POINT as output geometry type, and then click OK to create the output (Figure 5.26).

7. Open ArcScene and then add Denton.lasd, PowerLine3D, and Proximity5 to ArcScene. Change scene property and layer symbology if needed; then navigate the 3D scene (Figure 5.27).

8. Save your ArcMap and ArcScene projects, and Word document.

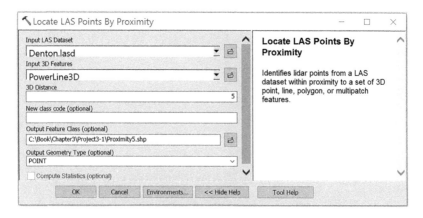

FIGURE 5.26   Locate LAS points by proximity tool.

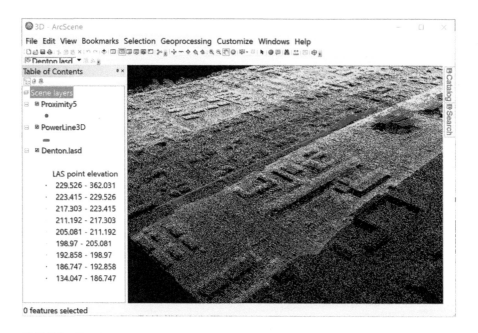

FIGURE 5.27   LiDAR dataset, 3D powerline, and proximity points in ArcScene.

PROJECT 5.2:   SMALL-AREA POPULATION ESTIMATION USING COUNT, AREA, AND VOLUME OF RESIDENTIAL BUILDINGS EXTRACTED FROM LiDAR DATA IN DENTON, TX, USA

1. Introduction

A small area can be defined as a single or aggregated sub-county area such as census tracts, block groups, and blocks (Dong et al. 2010). In this project, a 2009 LiDAR dataset acquired in the city of Denton, TX, USA and a 2010 census block shapefile (censusblocks.shp) with census data are used for small-area population estimation. Three variables (count, area, and volume of buildings) are extracted from LiDAR data and employed separately for population estimation. The basic idea is to create OLS regression equations using the three individual variables (x) and the 2010 census data (y) for selected census blocks (training samples). Once the OLS equations are established from the training samples, the total number, total area, and total volume of buildings in all census blocks are used to calculate estimated total population. Finally, values of estimated total population are compared with the 2010 census data to calculate the accuracy of population estimation.

2. Data

The LiDAR data was acquired by the Texas Natural Resources Information System (TNRIS) through the High Priority Imagery and Data Sets (HPIDS) contract in 2009, with a point density of about 1 point/m². A tile (TNRIS_2009_1.0M_339756_3_d.las) of about 2.9 km × 3.5 km was selected and renamed "Denton2009.las". The horizontal coordinate system for the LiDAR point clouds is NAD83_UTM_zone_14N. A 2010 census blocks shapefile "censusblocks.shp" (and seven other files with different extensions) is also available. The data for this project can be downloaded from http://geography.unt.edu/~pdong/LiDAR/Chapter5/Project5.2/ (right click each file to download the files).

3. Project Steps

   1. Open an empty Word document so that you can copy any results from the following steps to the document. To copy the whole screen to your Word document, press the PrtSc (print screen) key on your keyboard, then open your Word document and click the "Paste" button or press Ctrl+V to paste the content into your document. To copy an active window to your Word document, press Alt+PrtSc, then paste the content into your document.

   2. Open ArcMap, and load the Spatial Analyst Extension.

   3. Create an LAS dataset. Open ArcToolbox → Data Management Tools → LAS Dataset → Create LAS Dataset, specify the input file "Denton2009.las" and the output LAS dataset "Denton2009.lasd," and then click OK to create the LAS dataset. By default, the LAS dataset is automatically added to ArcMap. Users can zoom in to see the point elevation in different colors.

   4. Create a DTM. Neither DTM nor DSM are made from all LiDAR points; therefore, point filters should be defined before creating a DTM or DSM. Right click the LAS dataset created in Step 3 and open its properties. In the Filter tab of the properties, check "Ground" and "Water" under "Classification Codes", then click OK. (Note: The filter can also be defined using the "Ground" option in the predefined settings in the properties form, and the results can be slightly different).

      Once a filter is defined, the LAS dataset can be converted to a DTM raster using ArcToolbox → Conversion Tools → To Raster → LAS

Dataset to Raster, by setting the following parameters: Value Field: ELEVATION; Interpolation Type: Binning; Cell Assignment Type: AVERAGE; Void Fill Method: NATURAL_NEIGHBOR; Output Data Type: FLOAT; Sampling Type: CELLSIZE; Sampling Value: 1; and Z Factor: 1. The DTM is shown in Figure 5.28. Note: You should select the input LAS dataset from the drop-down list because the filter was defined through the layer properties form. If you use the browse button to select a LAS dataset as input, all the data points in the LAS files it references will be processed, and the defined filter will not be used.

5. Create a DSM. Similar to Step 4, a filter for a DSM should be defined. In the Filter tab of the LAS dataset properties, check "Ground", "Building", and "Water" under "Classification Codes", then click OK. Other classes are not selected when defining the filter for the DSM because the purpose of the project is to extract buildings for population estimation. Using the same parameters as in Step 4, a DSM can be created (Figure 5.29).

6. Create a DHM by subtracting DTM from DSM. Theoretically, a DHM is created as if the objects such as trees and building on the earth's surface are put on a flat surface, therefore the values of a DHM change from 0 to the maximum height of the objects. In reality, the minimum value of a DHM can be less than 0 due to the errors induced in LiDAR data collection and processing. To create a DHM, open ArcToolbox, then select Spatial Analyst Tools → Map Algebra → Raster Calculator. An output DHM raster "dhm" can be created by subtracting the DTM from the DSM. The DHM is shown in Figure 5.30.

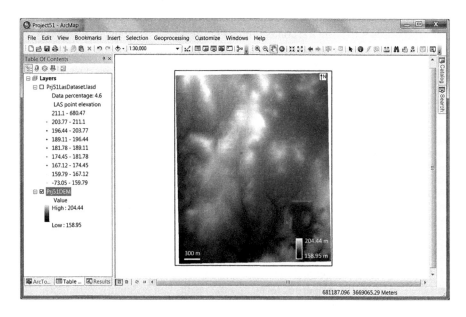

FIGURE 5.28 DTM created using LAS dataset to raster conversion.

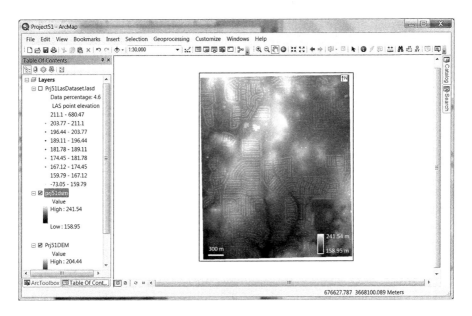

FIGURE 5.29   DSM created using LAS dataset to raster conversion.

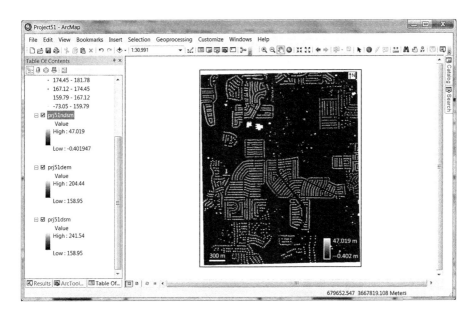

FIGURE 5.30   Digital height model (DHM).

7. Separate residential buildings from the DHM. Based on the Manufactured Home Construction and Safety Standards published by the Texas Department of Housing & Community Affairs (TDHCA 2009), a threshold of 2.2 m was used to separate residential buildings

from other ground objects. Cell values (heights) greater than 2.2 m in the DHM are saved in the new raster "dhm_bldg", whereas other cells are set to zero. This can be done using the conditional "Con()" function in the raster calculator. The expression is Con("dhm" > 2.2, "dhm", 0), and the output raster is "dhm_bldg".

8. Create a binary mask for residential buildings extracted in Step 7. For automated counting of the number of residential buildings in each census block, several steps are needed for further processing of the residential buildings extracted in Step 7. In this step, a binary mask "bldg_mask" will be created from the raster "dhm_bldg". If the cell value of the raster "dhm_bldg" is greater than 0, the corresponding output value in "bldg_mask" is 1; otherwise, the output is 0. This can be done using the conditional "Con()" function in the raster calculator. The expression is Con("dhm_bldg" > 0, 1, 0), and the output raster is "bldg_mask".

9. Remove noise in the binary mask for residential buildings. Open ArcToolbox → Spatial Analyst Tools → Neighborhood → Focal Statistics, use "bldg_mask" as the input raster, "bldg_mask2" as the output raster, "Rectangle" as the neighborhood, 3 as the height and width, "Cell" as the unit, MEDIAN as the statistics type, and click OK. Figure 5.31 shows the comparison between the two rasters "bldg_mask" (Figure 5.31A) and "bldg_mask2" (Figure 5.31B). It can be seen that holes and isolated cells in "bldg_mask" have been removed in "bldg_mask2".

10. Create building footprints using raster to polygon conversion. Using Raster to Polygon conversion (ArcToolbox → Conversion Tools → From Raster → Raster to Polygon), input raster "bldg_mask2" can be converted into a polygon shapefile "bldg_poly.shp" using VALUE as the conversion field. Note that the building boundaries can be simplified or even distorted if the "Simply polygon" option is checked, but that does not affect our results because we are interested in counting the number of polygons (buildings) instead of delineating actual building footprints. Figure 5.32 shows converted building polygons.

At this point, the following datasets have been created: (1) A building polygon shapefile (bldg_poly.shp) which can be used to obtain the

(A)   (B)

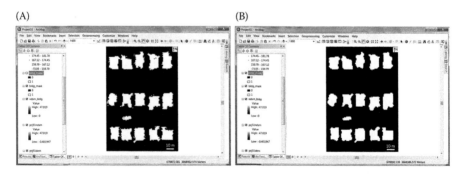

FIGURE 5.31 Comparison between the two rasters "bldg_mask" (A) and "bldg_mask2" (B).

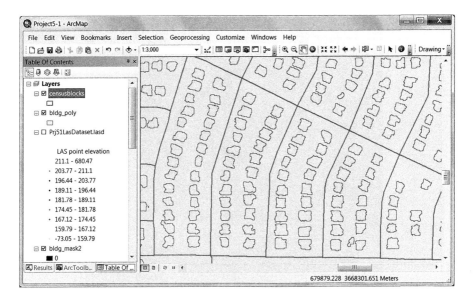

FIGURE 5.32 Converted building polygons (bldg_poly) within census blocks.

number of buildings in each census block, (2) A binary raster for buildings (bldg_mask2) which can be used to get the area of buildings in each census block, and (3) a building height raster "dhm_bldg" which can be used to obtain the building volume in each census block. Using these variables from selected census blocks as x, and the corresponding 2010 population (available in the census block shapefile) as y, OLS regression equations can be obtained to estimate total population in the study area.

11. Select census blocks as training samples. A total of 30 census blocks are randomly selected as training sample and saved as a new shapefile "blocksamples.shp" (Figure 5.33). In the following steps, the population, building count, building area, and building volume of the sample blocks are used to build regression models.

12. Obtain number of buildings in each census block using spatial join. Open ArcToolbox → Analysis Tools → Overlay → Spatial Join (Figure 5.34), use censusblocks as target features, bldg_poly as join features, sp-join.shp as output feature class, and CONTAINS as match option to create spatial join between census blocks and building polygons. In the attribute table of the output feature class, the field "Join_Count" represents the number of buildings in each census block. Note that the default match option is INTERSECT, which will produce an extra count in the "Join_Count" field. Using CONTAINS as match option will produce exact number of buildings in each census block.

13. Obtain building area in each census block using Zonal Statistics as Table. Open ArcToolbox → Spatial Analyst Tools → Zonal → Zonal Statistics as Table, use blocksamples.shp as the feature zone data, FID as the zone field, and the binary raster "bldg_mask2" as the input value

FIGURE 5.33 Thirty census blocks selected as training samples.

raster, select "SUM" as the statistics type, and click OK to produce zonal statistics in the output table. Since the binary raster "bldg_mask2" has cell values of 0 for non-buildings and 1 for buildings, the statistics type SUM will report the total number of building cells in each census block in blocksamples.shp. The SUM field in the output table represents not only the number of building cells but also the area of building cells in each census block because the cell size is 1 m × 1 m. For more information on the Zonal Statistics as Table tool, refer to Project 2.1.

14. Obtain building volume in each census block using Zonal Statistics as Table. Similar to Step 13, building volume in each sample census block can be obtained from ArcToolbox → Spatial Analyst Tools → Zonal → Zonal Statistics as Table. Select blocksamples.shp as the feature zone data, FID as the zone field, and the binary raster "dhm_bldg" as the input value raster, select "SUM" as the statistics type, and click OK to produce zonal statistics in the output table.

15. Create regression models. The results of Steps 9–12 are summarized in Table 5.3. The table can be exported as a text file, and imported to Excel to create regression models (Figures 5.35 through 5.37).

Table 5.3. Population, building count, building area, and building volume in 30 sample census blocks.

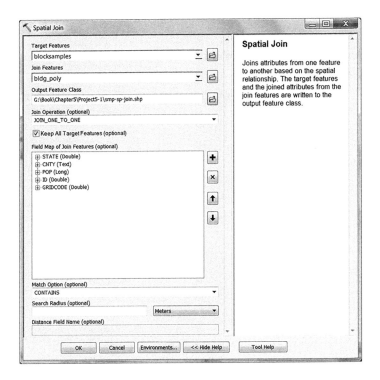

FIGURE 5.34   Spatial join tool in ArcGIS.

16. Obtaining total building count, total building area, and total building volume in all census blocks, and calculating estimated population.

   Note: The total building count, total building area, and building volume should be obtained from the 141 census blocks, not from the entire rasters because some areas in the rasters are not covered by the census blocks in this project.

   A total of 2650 buildings are selected in 141 census blocks, and the estimated population is 7792 based on the regression model in Figure 5.35. The total building area (668,775 m^2) and total building volume (3,208,273.81 m^3) can be obtained using the same methods as in Steps 13 and 14, but with "censusblocks.shp" as the feature zone data. Based on the regression models in Figures 5.36 and 5.37, the population estimates are 8629 and 8347. Compared with total population (8235) of the 141 census blocks in the study area (Census 2010 data), the relative errors of population estimation using building count, building area, and building volume are 5.38%, 4.78%, and 1.36%, respectively. As can be seen from the results, it is possible to achieve over 95% accuracy for small-area population estimation using LiDAR data. It appears that building volume provides the best population estimates in this project. However, the conclusion could be different if a different set of training blocks are selected for constructing regression equations.

TABLE 5.3

Population, Building Count, Building Area, and Building Volume in 30 Sample Census Blocks

FID	Population	Bldg_Count	Bldg_Area	Bldg_Volume
0	94	24	7,595.00	43,251.12
1	57	23	4,913.00	25,288.32
2	42	2D	3,508.00	12,617.39
3	92	46	7,985.00	29,809.72
4	25	10	3,518.00	17,568.79
5	54	15	4,285.00	23,646.96
6	28	12	2,450.00	11,367.02
7	63	19	4,512.00	18,945.73
8	99	35	7,013.00	37,383.58
9	85	30	7,060.00	32,216.75
10	141	29	10,874.00	54,907.82
11	121	35	9,922.00	45,621.58
12	52	15	3,640.00	16,160.32
13	41	14	3,356.00	15,305.04
14	134	37	8,632.00	38,827.24!
15	31	9	2,092.00	9,935.32
16	19	7	1,313.00	5,993.33
17	158	41	10,007.00	51,559.16
18	42	15	2,598.00	11,890.73
19	95	35	7,696.00	33,419.58
20	70	23	4,062.00	14,449.14
21	131	50	11,002.00	44,940.65
22	44	16	3,304.00	14,602.13
23	47	14	2,548.00	11,777.06
24	35	11	2,470.00	10,485.93
25	93	30	6,859.00	32,366.75
26	36	13	2,648.00	12,661.37
27	117	40	9,545.00	47,094.28
28	47	16	3,812.00	16,441.43
29	20	7	1,604.00	7,237.89

17. Save your ArcMap project and Word document.
18. Questions: (1) The relationship between a dependent variable (such as population) and one or more independent variables (such as building count, building area, and building volume) might vary geographically. How can you take into account geographical differences when estimating population using the independent variables? (2) Identification of buildings using LiDAR data can be affected by tree canopy. What options do you have to reduce the influence of tree canopy on building detection?

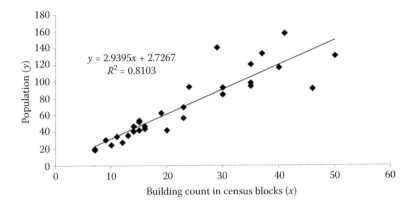

FIGURE 5.35 Regression model for population derived from building count in census blocks.

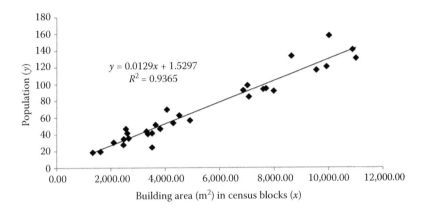

FIGURE 5.36 Regression model for population derived from building area in census blocks.

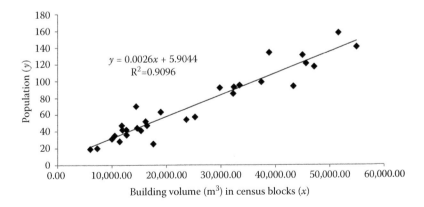

FIGURE 5.37 Regression model for population derived from building volume in census blocks.

REFERENCES

Adams, B.J., 2004. Improved disaster management through post-earthquake building damage assessment using multitemporal satellite imagery. In: *Proceedings of the ISPRS 20th Congress*, Volume 35, July 12–23, 2004, Istanbul, Turkey.

Adams, B.J., Huyck, C.K., Mansouri, B., Eguchi, R.T., and Shinozuka, M., 2004. Application of high-resolution optical satellite imagery for post-earthquake damage assessment: the 2003 Boumerdes (Algeria) and Bam (Iran) earthquakes. In MCEER Research and Accomplishments 2003–2004, pp. 173–186 (Buffalo, NY: MCEER).

Adams, B.J., Ghosh, S., Wabnitz, C., and Alder, J., 2005. Post-tsunami urban damage assessment in Thailand, using optical satellite imagery and the VIEWS™ field reconnaissance system. In: *Proceedings of the Conference on the 250th Anniversary of the 1755 Lisbon Earthquake*, November 1–4, 2005, Lisbon, Portugal.

Adams, B.J., Mansouri, B., and Huyck, C.K., 2006. Streamlining post-earthquake data collection and damage assessment in Bam, using VIEWS (Visualizing Impacts of Earthquake with Satellites), Earthquake Spectra Special Issue, 2003. Bam, Iran, Earthquake Reconnaissance Report, EERI, Oakland, CA.

Addink, E.A., Van Coillie, F.M.B., and De Jong, S.M., 2012. Introduction to the GEOBIA 2010 special issue: from pixels to geographic objects in remote sensing image analysis. *International Journal of Applied Earth Observation and Geoinformation*, 15: 1–6.

Ahmadi, S., Zoej, M.J., Ebadi, H., Moghaddam, H.A., and Mohammadzadeh, A., 2010. Automatic urban building boundary extraction from high resolution aerial images using an innovative model of active contours. *International Journal of Applied Earth Observation and Geoinformation*, 12: 150–157.

Alexander, C., Smith-Voysey, S., Jarvis, C., and Tansey, K., 2009. Integrating building footprints and LiDAR elevation data to classify roof structures and visualise buildings. *Computers, Environment and Urban Systems*, 33: 285–292.

Aoki, H., Matsuoka, M., and Yamazaki, F., 1998. Characteristics of satellite SAR images in the damaged areas due to the Hyogoken-Nanbu earthquake. In: *Proceedings of the 1998 Asian Conference on Remote Sensing*, November 16–20, Manila.

Aplin, P., and Smith, G.M., 2008. Advances in object-based image classification. *International Archives of the Photogrammetry, Remote Sensing and Spatial Information Sciences. ISPRS*, 37: 725–728.

Awrangjeb, M., and Fraser, C.S., 2014. Automatic segmentation of raw LIDAR data for extraction of building roofs. *Remote Sensing*, 6: 3716–1751.

Baltsavias, E. P., 2004. Object extraction and revision by image analysis using existing geodata and knowledge: Current status and steps towards operational systems. *ISPRS Journal of Photogrammetry and Remote Sensing*, 58: 129–151.

Baltsavias, E. P., Mason, S., and Stallmann, D., 1995. Use of DTMs/DSMs and orthoimages to support building extraction. In (A. Grün, O. Kübler, and P. Agouris, eds) *Automatic Extraction of Man-Made Objects from Aerial and Space Images*, Birkhäuser Verlag, Basel, Switzerland, pp. 199–210.

Barsi, A., 2004. Object detection using neural self-organisation. *The International Archives of the Photogrammetry, Remote Sensing and Spatial Information Sciences*, XXXV (B3): 366–371.

Bellman, C.J., and Shortis, M.R., 2004. A classification approach to finding buildings in large scale aerial photographs. *The International Archives of the Photogrammetry, Remote Sensing and Spatial Information Sciences*, XXXV (B3): 337–342.

Blaschke, T., 2010. Object based image analysis for remote sensing. *ISPRS Journal of Photogrammetry and Remote Sensing*, 65: 2–16.

Boyko, A., and Funkhouser, T., 2011. Extracting roads from dense point clouds in large scale urban environment. *ISPRS Journal of Photogrammetry and Remote Sensing*, 66: S2–S12.

Bracken, I., 1991. A surface model approach to small area population estimation. *Town Planning Review*, 62: 225–37.

Brenner, C., 2005. Building reconstruction from images and laser scanning. *International Journal of Applied Earth Observation and Geoinformation*, 6: 187–198.

Cai, Q., 2007. New techniques in small area population estimates by demographic characteristics. *Population Research and Policy Review*, 26: 203–218.

Coppin, P., Jonckheere, I., Nackaerts, K., Muys, B., and Lambin, E., 2004. Review article digital change detection methods in ecosystem monitoring: A review. *International Journal of Remote Sensing*, 25: 1565–1596.

Coxeter, H.S.M., 1969. Barycentric coordinates. In: *Introduction to Geometry* (2nd edition), Wiley, New York, pp. 216–221.

Cui, C., 2014. Impervious surface mapping using high-resolution images and LiDAR data in Denton, Texas. Master's thesis, Department of Geography, University of North Texas, 78 p.

Deer, P., 1995. Digital change detection techniques in remote sensing. Technical report, DSTO-TR-0169. Department of Defence, Australia, p. 52.

Dong, P., 1997. Implementation of mathematical morphological operations for spatial data processing. *Computers and Geosciences*, 23: 103–107.

Dong, P., and Guo, H.D., 2012. A framework for automated assessment of post-earthquake building damage using geospatial data. *International Journal of Remote Sensing*, 33: 81–100.

Dong, L., and Shan, J., 2013. A comprehensive review of earthquake-induced building damage detection with remote sensing techniques. *ISPRS Journal of Photogrammetry and Remote Sensing*, 84: 85–99.

Dong, P., Ramesh, S., and Nepali, A., 2010. Evaluation of small area population estimation using LiDAR, Landsat TM and parcel data. *International Journal of Remote Sensing*, 31: 5571–5586.

Dong, P., Zhong, R., and Yigit, A., 2018. Automated parcel-based building change detection using multitemporal airborne LiDAR data. *Surveying and Land Information Science*, (in press).

Elwood, S., 2008. Volunteered geographic information: Future research directions motivated by critical, participatory, and feminist GIS. *GeoJournal*, 72: 173–183.

Ferraz, A., Mallet, C., and Chehata, N., 2016. Large-scale road detection in forested mountainous areas using airborne topographic lidar data. *ISPRS Journal of Photogrammetry and Remote Sensing*, 112: 23–36.

Fischler, M.A., and Bolles, R.C., 1981. Random sample consensus: A paradigm for model fitting with applications to image analysis and automated cartography. *Communications of the ACM*, 24: 381–395.

Flanagin, A.J., and Metzger, M.J., 2008. The credibility of volunteered geographic information. *GeoJournal*, 72: 137–148.

Goodchild, M.F., 2007. Citizens as sensors: The world of volunteered geography. *GeoJournal*, 69: 211–221.

Goodchild, M.F., 2009. NeoGeography and the nature of geographic expertise. *Journal of Location Based Services*, 3: 82–96.

Gruen, A., 1998. TOBAGO—A semi-automated approach for the generation of 3-D building models. *ISPRS Journal of Photogrammetry and Remote Sensing*, 53: 108–118.

Guo, H., Li, X., and Zhang, L., 2009. Study of detecting method with advanced airborne and spaceborne synthetic aperture radar data for collapsed urban buildings from the Wenchuan earthquake. *Journal of Applied Remote Sensing*, 3: 2–19.

Haala, N., and Brenner, C., 1999. Extraction of buildings and trees in urban environments. *ISPRS Journal of Photogrammetry and Remote Sensing*, 54: 130–137.

Harvey, J.T., 2002, Population estimation models based on individual TM pixels. *Photogrammetric Engineering and Remote Sensing*, 68: 1181–1192.

Harvey, J.T., 2003. Population estimation at the pixel level: Developing the expectation maximization technique. In (V. Mesev, ed.) *Remotely Sensed Cities*, Taylor & Francis, London, pp. 181–205.

Hay, G.J., Castilla, G., Wulder, M.A., and Ruiz, J.R., 2005. An automated object-based approach for the multiscale image segmentation of forest scenes. *International Journal of Applied Earth Observation and Geoinformation*, 7: 339–359.

Hebel, M., Arens, M., and Stilla, U., 2013. Change detection in urban areas by object-based analysis and on-the-fly comparison of multi-view ALS data. *ISPRS Journal of Photogrammetry and Remote Sensing*, 86: 52–64.

Heipke, C., Mayer, H., Wiedemann, C., Sensing, R., and Jamet, O., 1997. Evaluation of automatic road extraction. *International Archives of Photogrammetry and Remote Sensing*, 32: 47–56.

Hodgson, M.E., Jensen, J.R., Tullis, J.A., Riordan, K.D., and Archer, C.M., 2003. Synergistic use of Lidar and color aerial photography for mapping urban parcel imperviousness. *Photogrammetric Engineering and Remote Sensing*, 69: 973–980.

Hussain, E., Ural, S., Kim, K., Fu, C.-S., and Shan, J., 2011. Building extraction and rubble mapping for city Port-au-Prince post-2010 earthquake with GeoEye-1 imagery and lidar data. *Photogrammetric Engineering and Remote Sensing*, 77: 1011–1023.

Hussain, M., Chen, D., Cheng, A., Wei, H., and Stanley, D., 2013. Change detection from remotely sensed images: From pixel-based to object-based approaches. *ISPRS Journal of Photogrammetry and Remote Sensing*, 80: 91–106.

Huyck, C.K., Mansouri, B., Eguchi, R.T., Houshmand, B., Castner, L.L. and Shinozuka, M., 2002. Earthquake damage detection algorithms using optical and ERS-SAR satellite data—Application to the August 17, 1999 Marmara, Turkey earthquake. In: *Proceedings of the 7th National US Conference on Earthquake Engineering*, Boston. 21–25 July, 2002.

Ioannidis, C., Psaltis, C., and Potsiou, C., 2009. Towards a strategy for control of suburban informal buildings through automatic change detection. *Computers, Environment and Urban Systems*, 33: 64–74.

Jang, K.H., and Jung, S.K., 2009. Practical modeling technique for large-scale 3D building models from ground images. *Pattern Recognition Letters*, 30: 861–869.

Jarosz, B., 2008. Using assessor parcel data to maintain housing unit counts for small area population estimates. In: (S. Murdock, and D. Swanson, eds) *Applied Demography in the 21st Century*, Springer, Dordrecht, The Netherlands, pp. 89–102.

Kabolizade, M., Ebadi, H., and Ahmadi, S., 2010. An improved snake model for automatic extraction of buildings from urban aerial images and LiDAR data. *Computers, Environment and Urban System*, 34: 435–441.

Kaya, S., Curran, P.J., and Llewellyn, G., 2005. Post-earthquake building collapse: A comparison of government statistics and estimates derived from SPOT HRVIR data. *International Journal of Remote Sensing*, 26: 2731–2740.

Kim, M., Madden, M., and Warner, T., 2008. Estimation of optimal image object size for the segmentation of forest stands with multispectral IKONOS imagery. In: (T. Blaschke, S. Lang, and G.J. Hay, eds) *Object-Based Image Analysis*, Springer, Berlin, Heidelberg, pp. 291–307.

Kumar, P., McElhinney, C.P., Lewis, P., and McCarthy, T., 2013. An automated algorithm for extracting road edges from terrestrial mobile LiDAR data. *ISPRS Journal of Photogrammetry and Remote Sensing*, 85: 44–55.

Landa, J., and Prochazka, D., 2014. Automatic road inventory using LiDAR. *Procedia Economics and Finance*, 12: 363–370.

Lein, J.K., 2012. *Object-Based Analysis, Environmental Sensing: Analytical Techniques for Earth Observation*, Springer, London, pp. 259–278.

Li, D., 2010. Remotely sensed images and GIS data fusion for automatic change detection. *International Journal of Image and Data Fusion*, 1: 99–108.

Li, G., and Weng, Q., 2005. Using Landsat ETM+ imagery to measure population density in Indianapolis, Indiana, USA. *Photogrammetric Engineering and Remote Sensing*, 71: 947–958.

Li, M., Cheng, L., Gong, J., Liu, Y., Chen, Z., Li, F., Chen, G., Chen, D., and Song, X., 2008. Post-earthquake assessment of building damage degree using LiDAR data and imagery. *Science in China Series E: Technological Sciences*, 51: 133–143.

Li, Y., Yong, B., Wu, H., An, R., and Xu, H., 2015. Road detection from airborne LiDAR point clouds adaptive for variability of intensity data. *Optik*, 126: 4292–4298.

Liu, C., Shi, B., Yang, X., Li, N., and Wu, H., 2013. Automatic buildings extraction from LiDAR data in urban area by neural oscillator network of visual cortex. *IEEE Journal of Selected Topics in Applied Earth Observations and Remote Sensing*, 6: 2008–2019.

Liu, W., Dong, P., Liu, S., and Liu, J., 2014. A rich Internet application for automated detection of road blockage in post-disaster scenarios. *IOP Conference Series: Earth and Environmental Science*, 18: 012124, doi: 10.1088/1755-1315/18/1/012124.

Lo, C.P., 1986a. Accuracy of population estimation from medium-scale aerial photography. *Photogrammetric Engineering and Remote Sensing*, 52: 1859–1869.

Lo, C.P., 1986b. *Applied Remote Sensing*, Longman, New York.Lo, C.P., 2003. Zone-based estimation of population and housing units from satellite generated land use/land cover maps. In: (V. Mesev, ed.) *Remotely Sensed Cities*, Taylor & Francis, London, pp. 157–180.

Lo, C.P., 2008. Population estimation using geographically weighted regression. *Journal GIScience and Remote Sensing*, 45: 131–148.

Lo, C.P., and Welch, R., 1977. Chinese urban population estimation. *Annals of the Association of American Geographers*, 67: 246–253.

Longley, P.A., 2002. Geographical information systems: Will developments in urban remote sensing and GIS lead to better urban geography? *Progress in Human Geography*, 26: 231–239.

Lu, D., Mausel, P., Brondízio, E., and Moran, E., 2004. Change detection techniques. *International Journal of Remote Sensing*, 25: 2365–2401.

Lu, D., Weng, Q., and Li, G., 2006. Residential population estimation using a remote sensing derived impervious surface approach. *International Journal of Remote Sensing*, 27: 3553–3570.

Maas, H.-G., and Vosselman, G., 1999. Two algorithms for extracting building models from raw laser altimetry data. *ISPRS Journal of Photogrammetry and Remote Sensing*, 54: 153–163.

Madhavan, B., Wang, C., Tanahashi, H., Hirayu, H., Niwa, Y., Yamamoto, K., Tachibana, K., and Sasagawa, T., 2006. A computer vision based approach for 3D building modelling of airborne laser scanner DSM data. *Computers, Environment and Urban Systems*, 30: 54–77.

Matsuoka, M., and Yamazaki, F., 2002. Application of the damage detection method using SAR intensity images to recent earthquakes. In: *Proceedings of the IGARSS*, Toronto. 24–28 June, 2002.

Mayer, H., 2008. Object extraction in photogrammetric computer vision. *ISPRS Journal of Photogrammetry and Remote Sensing*, 63: 213–222.

Mongus, D., Lukac, N., Obrul, D., and Žalik, B., 2013. Detection of planar points for building extraction from Lidar data based on differential morphological and attribute profiles. *ISPRS Annals of the Photogrammetry, Remote Sensing and Spatial Information Sciences*, Volume II-3/W1, 2013 VCM 2013—The ISPRS Workshop on 3D Virtual City Modeling, May 28, 2013, Regina, Canada.

Mouat, D.A., Mahin, G.G., and Lancaster, J., 1993. Remote sensing techniques in the analysis of change detection. *Geocarto International*, 8: 39–50.

Nixon, M., and Aguado, A.S., 2002. *Feature Extraction and Image Processing*, Newenes, Oxford, UK, pp. 220–243.

Oriot, H., 2003. Statistical snakes for building extraction from stereoscopic aerial images. *The International Archives of the Photogrammetry, Remote Sensing and Spatial Information Sciences*, 34 (.3/W 8): 65–70.

Orthuber, E., and Avbelj, J., 2015. 3D building reconstruction from Lidar point clouds by adaptive dual contouring. *ISPRS Annals of the Photogrammetry, Remote Sensing and Spatial Information Sciences*, Volume II-3/W4, 2015 PIA15+HRIGI15—Joint ISPRS conference 2015, 25–27 March, 2015, Munich, Germany.

Pesaresi, M., and Benediktsson, J.A., 2001. A new approach for the morphological segmentation of high-resolution satellite imagery. *IEEE Transaction on Geosciences and Remote Sensing*, 39: 309–320.

Platek, R., Rao, J., Sarndal, C., and Singh, M., 1987. *Small Area Statistics: An International Symposium*. John Wiley & Sons, New York.

Pu, S., and Vosselman, G., 2009. Knowledge based reconstruction of building models from terrestrial laser scanning data. *ISPRS Journal of Photogrammetry and Remote Sensing*, 64: 575–584.

Qiu, F., Woller, K.L., and Briggs, R., 2003. Modeling urban population growth from remotely sensed imagery and TIGER GIS road data. *Photogrammetric Engineering and Remote Sensing*, 69: 1031–1042.

Qiu, F., Sridharan, H., and Chun, Y., 2010. Spatial autoregressive model for population estimation at the census block level using LIDAR-derived building volume information. *Cartography and Geographic Information Science*, 37: 239–257.

Rottensteiner, F., Trinder, J., Clode, S., and Kubik, K., 2005. Using the Dempster–Shafer method for the fusion of LIDAR data and multi-spectral images for building detection. *Information Fusion*, 6: 283–300.

Saito, K., Spence, R.J., Going, C., and Markus, M., 2004. Using high-resolution satellite images for post-earthquake building damage assessment: A study following the 26.1.01 Gujarat earthquake. *Earthquake Spectra*, 20: 145–170.

Serra, J., 1983. *Image Analysis and Mathematical Morphology*, Academic Press, Orlando, FL.

Singh, A., 1989. Digital change detection techniques using remotely-sensed data. *International Journal of Remote Sensing*, 10: 989–1003.

Singh, K.K., Vogler, J.B., Shoemaker, D.A., and Meentemeyer, R.K., 2012. LiDAR-Landsat data fusion for large-area assessment of urban land cover: Balancing spatial resolution, data volume and mapping accuracy. *ISPRS Journal of Photogrammetry and Remote Sensing*, 74: 110–121.

Sithole, G., and Vosselman, G., 2004. Experimental comparison of filter algorithms for bare earth extraction from airborne laser scanning point clouds. *ISPRS Journal of Photogrammetry and Remote Sensing*, 59: 85–101.

Smith, S.K., and Lewis, B., 1980. Some new techniques for applying the housing unit method of local population estimation. *Demography*, 17: 323–339.

Smith, S.K., and Lewis, B., 1983. Some new techniques for applying the housing unit method of local population estimation: further evidence. *Demography*, 20: 407–413.

Smith, S.K., and Mandell, M., 1984. A comparison of population estimation methods: housing unit versus component II, ratio correlation and administrative records. *Journal of American Statistical Association*, 79: 282–289.

Smith, S.K., and Cody, S., 1994. 'Evaluating the housing unit method: A case study of 1990 population estimates in Florida. *Journal of the American Planning Association*, 60: 209–21.

Stilla, U., and Jurkiewicz, K., 1999. Reconstruction of building models from maps and laser altimeter data. In: (P. Agouris, and A. Stefanidis, eds) *Integrated Spatial Databases: Digital Images and GIS*, Springer, Berlin, pp. 34–46.

Stilla, U., Soergel, U., and Thoennessen, U., 2003. Potential and limits of InSAR data for building reconstruction in built-up areas. *ISPRS Journal of Photogrammetry and Remote Sensing*, 58: 113–123.

Stow, D., 2010. Geographic object-based image change analysis. In: (M.M. Fischer, and A. Getis, eds) *Handbook of Applied Spatial Analysis*, Springer, Berlin and Heidelberg, pp. 565–582.

Sugihara, K., and Hayashi, Y., 2008. Automatic generation of 3D building models with multiple roofs. *Tsinghua Science and Technology*, 13: 368–374.

Susaki, J., 2013. Knowledge-based modeling of buildings in dense urban areas by combining airborne LiDAR data and aerial images. *Remote Sensing*, 5: 5944–5963.

Sutton, P., Roberts, D., Elvidge, C.D., and Meij, H., 1997. A comparison of nighttime satellite imagery and population density for the continental United States. *Photogrammetric Engineering and Remote Sensing*, 63: 1303–1313.

Sutton, P., Roberts, D., Elvidge, C.D., and Baugh, K., 2001. Census from heaven: An estimate of the global human population using night-time satellite imagery. *International Journal of Remote Sensing*, 22: 3061–3076.

Suveg, I., and Vosselman, G., 2004. Reconstruction of 3D building models from aerial images and maps. *ISPRS Journal of Photogrammetry and Remote Sensing*, 58: 202–224.

Teo, T.-A., and Shi, T.-Y., 2012. Lidar-based change detection and change type determination in urban areas. *International Journal of Remote Sensing*, 34: 968–981.

Texas Department of Housing & Community Affairs (TDHCA), 2009. Manufactured home construction and safety standards. http://www.tdhca.state.tx.us/mh/ (last Accessed on 14 September, 2009).

Tovari, D., and Pfeifer, N., 2005. Segmentation based robust interpolation—A new approach to laser data filtering. *The International Archives of the Photogrammetry, Remote Sensing and Spatial Information Sciences*, XXXVI (3/W19): 79–84.

Trinder, J., Salah, M., Shaker, A., Hamed, M., and ELsagheer, A., 2010. Combining statistical and neural classifiers using Dempster-Shafer theory of evidence for improved building detection. *The 15Th Australian Remote Sensing and Photogrammetry Conference*. Alice Springs, September 13–17, 2010.

Verma, R., Basavarajappa, K., and Bender, R., 1984. Estimation of local area population: An international comparison. In: *Proceedings of the Social Statistics Section*, American Statistical Association, Alexandria, VA, pp. 324–329.

Vu, T.T., Yamazaki, F., and Matsuoka, M., 2009. Multi-scale solution for building extraction from LiDAR and image data. *International Journal of Applied Earth Observation and Geoinformation*, 11: 281–289.

Wang, R., 2013. 3D building modeling using images and LiDAR: A review. *International Journal of Image and Data Fusion*, 4: 273–292.

Wang, C., Zhang, H., Wu, F., Zhang, B., Tang, X., Wu, H., Wen, X., and Yan, D., 2009. Disaster phenomena of Wenchuan earthquake in high resolution airborne synthetic aperture radar images. *Journal of Applied Remote Sensing*, 3: 20–35.

Weidner, U., and Förstner, W., 1995. Towards automatic building extraction from high-resolution digital elevation models. *ISPRS Journal of Photogrammetry and Remote Sensing*, 50: 38–49.

Welch, R., and Zupko, S., 1980. Urbanized area energy utilization patterns from DMSP data. *Photogrammetric Engineering and Remote Sensing*, 46: 1107–1121.

Wolter, K., and Causey, B., 1991. Evaluation of procedures for improving population estimates for small areas. *Journal of the American Statistical Association*, 86: 278–284.

Wu, C., 2004. Normalized spectral mixture analysis for monitoring urban composition using ETM+ imagery. *Remote Sensing of Environment*, 93: 480–492.

Wu, C., and Murray, A.T., 2003. Estimating impervious surface distribution by spectral mixture analysis. *Remote Sensing of Environment*, 84: 493–505.

Wu, B., Yang, J., Zhao, Z., Meng, Y., Yue, A., Chen, J., He, D., Liu, X., and Liu, S., 2014. Parcel-based change detection in land-use maps by adopting the holistic feature. *IEEE Journal of Selected Topics in Applied Earth Observation and Remote Sensing*, 7: 3482–3490.

Xiao, Y., Wang, C., Li, J., Zhang, W., Xi, X., Wang, C., and Dong, P., 2014. Building segmentation and modeling from airborne LiDAR data. *International Journal of Digital Earth*, 8: 694–709.

Xie, Y., Weng, A., and Weng, Q., 2015. Population estimation of urban residential communities using remotely sensed morphologic data. *IEEE Geoscience and Remote Sensing Letters*, 12: 1111–1115.

Yan, W.Y., Shaker, A., and El-Ashmawy, N., 2015a. Urban land cover classification using airborne LiDAR data: A review. *Remote Sensing of Environment*, 158: 295–310.

Yan, J., Zhang, K., Zhang, C., Chen, S.-C., and Narasimhan, G., 2015b. Automatic construction of 3-D building model from airborne LiDAR data through 2-D snake algorithm. *IEEE Transactions on Geoscience and Remote Sensing*, 53: 3–14.

Yusuf, Y., Matsuoka, M., and Yamazaki, F., 2002, Detection of building damage due to the 2001 Gujrat, India Earthquake, using satellite remote sensing. In: *Proceedings of the 7th National US Conference on Earthquake Engineering*, Boston. July 21–25, 2002.

Zavodny, A.G., 2012. Change detection in Lidar scans of urban environments. PhD dissertation, Graduate Program in Computer Science and Engineering, University of Notre Dame, Notre Dame, IN, 134 p.

Zhang, L., Xu, T., and Zhang, J., 2012. Building extraction based on multiscale segmentation. In: *5th International Congress on Image and Signal Processing (CISP)*, October 16–18, 2012, pp. 657–661, doi: 10.1109/CISP.2012.6470035.

Zhao, Y., Ovando-Montejo, G.A., Frazier, A.E., Mathews, A.J., Flynn, K.C., and Ellis, E.A., 2017. Estimating work and home population using lidar-derived building volumes. *International Journal of Remote Sensing*, 38: 1180–1196.

6 LiDAR for Geoscience Applications

6.1 INTRODUCTION

The unique capability of LiDAR in providing highly accurate x, y, and z coordinates of ground points makes it an ideal data source for studying features on the Earth's solid surface (Dong 2012). Research literature in geoscience applications of LiDAR can be summarized in six major fields: (1) changes in geomorphic surfaces, including fundamental topographic signatures (Perron et al. 2009), alluvial fan formative processes and debris flow deposits (Staley et al. 2006, Volker et al. 2007), volumetric changes of coastal dunes and beach erosion (Woolard and Colby 2002, Mitasova et al. 2004, Saye et al. 2005, Richter et al. 2011), changes in desert sand dunes (Ewing and Kocurek 2010a, 2010b, Reitz et al. 2010, Baitis et al. 2014, Dong 2015, Ewing et al. 2015), changes in glaciers/ice sheets and glacial sediment redistribution (Krabill et al. 1995, Irvine-Fynn et al. 2011), and lava flow dynamics and rheology (Tarquini and Favalli 2011, Jessop et al. 2012, Tarquini et al. 2012); (2) surface hydrology and flood models (Cavalli et al. 2008, Jones et al. 2008, Vianello et al. 2009, Fewtrell et al. 2011, Sampson et al. 2012); (3) tectonic geomorphology (Cunningham et al. 2006, Kondo et al. 2008, Arrowsmith and Zielke 2009, Begg and Mouslopoulou 2010, Howle et al. 2012, Dong 2014); (4) lithological mapping (Grebby et al. 2010, 2011); (5) rock mass structural analysis (Gigli and Casagli 2011, Lato and Vöge 2012, Lato et al. 2013); and (6) natural hazards, such as landslides, debris flows, and earthquake damage (Glenn et al. 2006, Schulz 2007, Bull et al. 2010, Lan et al. 2010, Dong and Guo 2012, Liu et al. 2012).

This chapter introduces LiDAR applications in the study of six major features/phenomena in geosciences: (1) Aeolian landforms (coastal dunes and desert dunes), (2) fluvial landforms (alluvial fans and terraces), (3) surface hydrology (watersheds and snow depth), (4) volcanic and impact landforms (volcanic cones and craters, lava flows, and impact craters), (5) tectonic landforms (linear and planar geomorphic markers), and (6) rocks and geologic structures. Finally, two step-by-step projects in ArcGIS are included at the end of the chapter to demonstrate measurement of dune migration using multi-temporal LiDAR data collected in the White Sands Dune Field (WSDF), NM, USA, and trend surface analysis of simple folds using LiDAR data collected in Raplee Ridge, UT, USA.

6.2 AEOLIAN LANDFORMS

Aeolian sand dunes are a major component of Aeolian landforms, and one of the most dynamic landforms on Earth and some other planets such as Mars and Venus and the moon Titan (Fenton 2006, Hugenholtz et al. 2007, Bourke et al. 2010).

Understanding how sand dunes form and change has long been a research topic in Earth and planetary surface processes (e.g., Bagnold 1941, Wasson and Hyde 1983, Lancaster 1995, Rubin and Hesp 2009, Bridges et al. 2012). In the 1970s and 1980s, many single-dune studies were carried out to understand the basic controls on the form of individual dunes (Livingstone et al. 2007). The rapid development in data collection and processing technology in the 1980s and 1990s led to more sophisticated studies of single dunes. Since 2000 there has been a shift in sand dune research focus from studying single dunes to studying dunes as complex systems (Livingstone et al. 2007). In addition to numerous field studies around the world (e.g., Rubin 1990, Ha et al. 1999, Dong et al. 2000, 2004, Elbelrhiti et al. 2005, Ewing and Kocurek 2010a, Zhang et al. 2012a), many other methods have been developed for sand dune studies, including cellular automaton models (Narteau et al. 2009, Zhang et al. 2010, Barrio-Parra and Rodríguez-Santalla 2014), numerical models (Alhajraf 2004, Hersen 2004, Zhang et al. 2012b, Araújo et al., 2013), flume experiments (Taniguchi et al. 2012), landscape-scale experiments (Ping et al. 2014), and remote sensing methods (e.g., Hunter et al. 1983, Gay 1999, Jimenez et al. 1999, Bailey and Bristow 2004). These studies have improved general understanding of sand dunes. A review of research progress in geomorphology of desert sand dunes can be found in the work of Livingstone et al. (2007).

In comparison with traditional remote sensing techniques, LiDAR has provided unprecedented datasets for sand dune studies, mostly in the form of high-resolution and high-accuracy digital elevation models (DEMs). Early LiDAR-based sand dune studies focused on morphometry and evolution of coastal dunes. For example, Woolard and Colby (2002) used multi-temporal LiDAR data acquired in 1996 and 1997 to obtain volumetric changes of coastal dunes at Cape Hatteras, NC, USA; Mitasova et al. (2004) used annual LiDAR data from 1997 to 2000, and global positioning system measurements, to study dune activity changes at Jockey's Ridge, NC, USA; and Saye et al. (2005) employed LiDAR data to investigate the relationships between frontal dune morphology and beach/near-shore morphology at five sites in England and Wales. Figure 6.1 shows perspective views of LiDAR point clouds

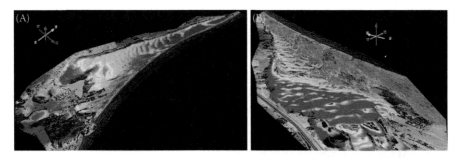

FIGURE 6.1 Perspective views of LiDAR point clouds collected on October 18, 2010 for two coastal dune fields—Vila Nova (A) and Garopaba (B) in southern Brazil. (Data provided by FAPESP grant 2009/17675-5, Grohmann, C.H., and Sawakuchi, A.O., *Geomorphology*, 180–181, 130–136, 2013.)

collected on October 18, 2010, for two coastal dune fields (Vila Nova and Garopaba) in southern Brazil.

Since 2010, several studies have been conducted for desert dunes. Reitz et al. (2010) used LiDAR data, acquired in June 2007 and June 2008, to study barchan-parabolic dune pattern transition from vegetation stability threshold in the WSDF, NM, USA. Based on manual digitizing of dune crestlines, Ewing and Kocurek (2010b) studied seven types of dune interactions at WSDF using digitally scanned aerial photographs from 1963 and 1985, digital orthophoto quarter quadrangles from 1996, 2003 and 2005, and a DEM with 1-m spatial resolution generated from a June 2007 airborne LiDAR survey. Baitis et al. (2014) used a LiDAR-derived DEM of a representative portion of WSDF to characterize dune-field parameters. Using LiDAR-derived DEMs, Ewing et al. (2015) investigated bedform patterns and processes that coexist in dune fields and that can be used to interpret environmental and climatic conditions in WSDF.

Dong (2015) proposed a new approach named Pairs of Source and Target Points (PSTP) for automated measurement of dune migration directions and migration rates using multi-temporal LiDAR data collected on January 24, 2009 (with a point density of 4.19 points m^{-2}) and June 6, 2010 (with a point density of 4.19 points m^{-2}) in WSDF (Figure 6.2). The theoretical foundation of the PSTP method is that sand avalanching and slumping events caused by gravity occur in the inclination direction of the slip face (Bagnold 1941). The centerlines of old slip faces are referred to as source lines, and the centerlines of new slip faces are referred to as target lines.

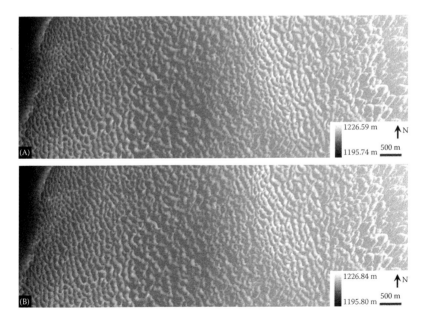

FIGURE 6.2   Digital elevation models (DEM) with 1 m × 1 m cell size created from multi-temporal LiDAR point clouds covering the study area of 2.4 km × 9 km in White Sands, NM (USA). (A) DEM for January 24, 2009, and (B) DEM for June 6, 2010.

The basic concept of the PSTP method is explained as follows: For any point t (target point) on a target line (Figure 6.3), there might be a nearest point s (source point) on a source line (or the extension of the source line) within a certain search radius, and the vector **A** is normal to the source line at point s. The length of the vector **A** is the dune migration distance, and the direction of the source point s relative to the target

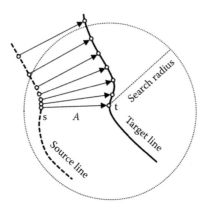

FIGURE 6.3 The concept of Pairs of Source and Target Points (PSTP). Source point s is the nearest point for target point t, and vector **A** is normal to the source line at point s.

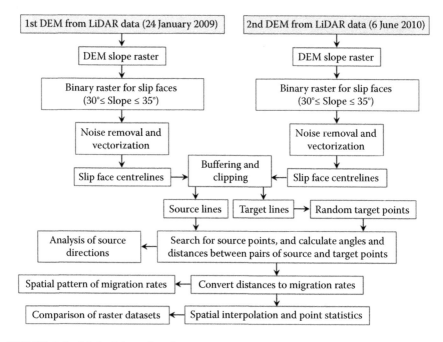

FIGURE 6.4 Methodology flowchart showing major steps of automated measurement of sand dune migration rates using multi-temporal LiDAR data.

point t is called source direction, counting clockwise from 0° (north) to 360°. Source directions do not necessarily follow the prevailing wind direction, but may reflect the prevailing wind direction statistically. Random points can be generated on the target line; thereby pairs of source and target points can be identified for automated calculation of migration distance and source direction. The major steps for automated measurement of dune migration are described in the flowchart in Figure 6.4. Some results are shown in Figures 6.5 through 6.8.

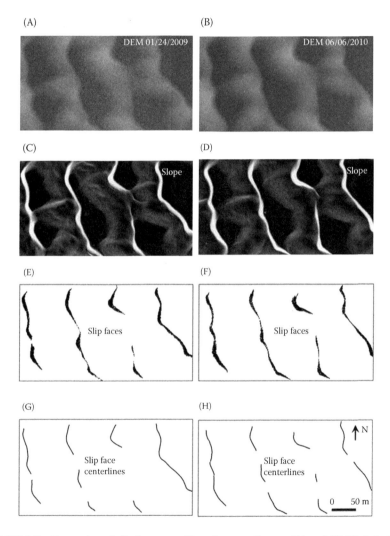

FIGURE 6.5 Extraction of slip face centrelines for a small area. (A) and (B) Digital elevation models for January 24, 2009, and June 6, 2010; (C) and (D) slope rasters derived from DEMs; (E) and (F) binary rasters for slip faces extracted from DEM slopes by setting values between 30° and 35° to 1, and other values to 0 and (G) and (H) vectorized centrelines for slip faces.

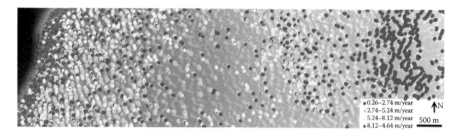

FIGURE 6.6 Sand dune migration rates of 5936 target points draped over LiDAR-derived DEM for June 6, 2010.

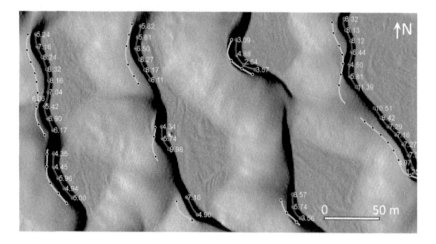

FIGURE 6.7 Sand dune migration rates (m year^{-1}) displayed as labels for target points for the small area in Figure 6.5. Source lines (green), target lines (red), and pairs of source and target points (black) are also shown with a background of a shaded DEM for June 6, 2010.

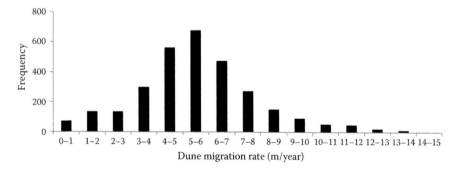

FIGURE 6.8 Histogram of dune migration rates (m year^{-1}) for 3025 target points with source direction in the range of 225°–285°.

6.3   FLUVIAL LANDFORMS

Alluvial fans have long been used as important records of quaternary climate and tectonics in arid and semiarid areas (Davis 1905, Wallace 1977, Nichols et al. 2006). Figure 6.9 shows 30-m resolution Landsat Thematic Mapper (TM) image, color composite of principal components of Landsat TM image bands, and a 1-m resolution hillshaded DEM derived from airborne LiDAR data for alluvial fans in Death Valley, California, USA. While optical images such as Landsat TM can provide spectral information of the alluvial fans, surface roughness is usually obtained from radar images (e.g., Sabins 1987). LiDAR provides a new data source for mapping surface roughness of alluvial fans using parameters such as slope, curvature, and aspect derived from high-resolution DEMs, as reported in Regmi et al. (2014).

Terraces can be formed in different geologic and environmental settings (Shaw 1911, Easterbrook 1999). High-resolution and high-accuracy DEMs derived from LiDAR data can be very useful in studying terrace formation and abandonment. Figure 6.11 shows a LiDAR-derived DEM of a section of South Fork Eel River, CA, USA. Two profiles across the terraces and river bed are shown in Figure 6.11, which shows that LiDAR data can reveal terraces, as well as micro-topographic variations on the terraces.

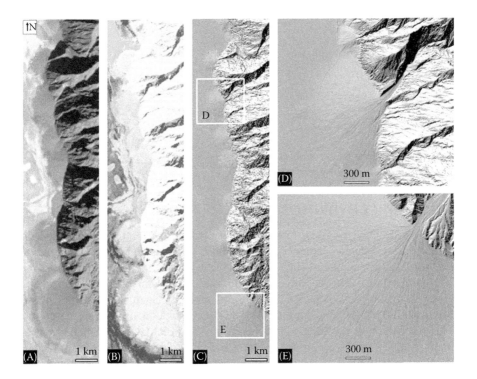

FIGURE 6.9   Landsat TM images and hillshaded DEM derived from LiDAR data for alluvial fans in Death Valley, California (USA). (A) Landsat TM image; (B) Color composite of principal components of Landsat TM bands; (C) Hillshaded DEM created from LiDAR data and (D) and (E) Sub-windows for two alluvial fans.

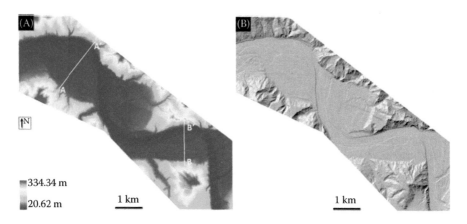

FIGURE 6.10 LiDAR-derived DEM of a section of South Fork Eel River, CA (USA). (A) 1-m resolution DEM; A-A′ and B-B′ are profile locations. Profiles are shown in Figure 6.11 and (B) hillshaded DEM.

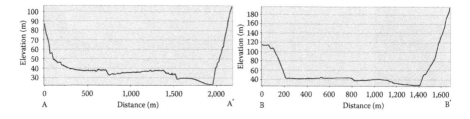

FIGURE 6.11 Topographic profiles derived from Figure 6.10A.

6.4 SURFACE HYDROLOGY

Surface hydrology is closely related to precipitation, evaporation, river channels, watersheds, and human activities, among others. This section provides examples of LiDAR data for mapping watersheds and snow depth distribution.

High-resolution DEMs are needed for accurate delineation of watersheds. Figure 6.12 shows a comparison of a 90-m resolution Shuttle Radar Topography Mission (SRTM) DEM (Figure 6.12A), a 30-m resolution Advanced Spaceborne Thermal Emission and Reflection Radiometer (ASTER) Global Digital Elevation Map (GDEM) DEM (Figure 6.12B), and a 1-m resolution LiDAR-derived DEM (Figure 6.12C) for a watershed in the Sawtooth National Forest located 27 km west of Stanley, ID, USA. Figures 6.12C through E are shaded DEMs of Figures 6.12A through C. As can be seen, the LiDAR-derived DEM provides the best data quality, which allows for more accurate calculation of flow directions and delineation of watersheds.

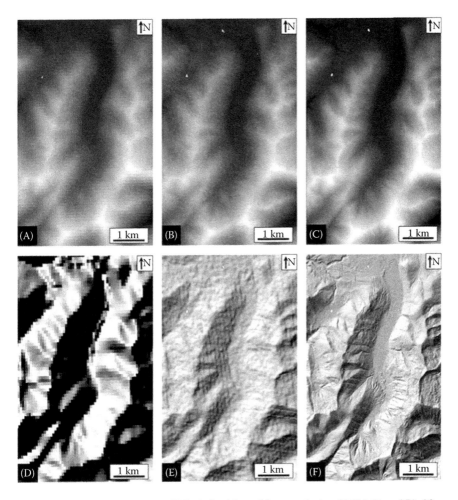

FIGURE 6.12 Comparison of DEMs derived from 90-m resolution SRTM (A and D), 30-m resolution ASTER GDEM (B and E), and 1-m resolution LiDAR-derived DEM (C and F) for a watershed in the Sawtooth National Forest, ID.

The heterogeneous distribution of snow cover in mountain watersheds can be caused by the variability in meteorological, topographical, and vegetative controls, among other factors (Dye and Tucker 2003, Tong et al. 2009, She et al. 2015). The spatial distribution of snow depth can be obtained from LiDAR data collected in snow-on and snow-off conditions. Figure 6.13 shows original and shaded DEMs created from snow-on and snow-off LiDAR data for an area around the Redondo Peak near the upper Jerez River basin in New Mexico, USA. The snow-on LiDAR data was collected during the peak snowpack season (March–April) in 2010 with a point density of 9.08 points/m^2, and the snow-off LiDAR data was collected during June and July of 2010 with a point density of 9.68 points/m^2.

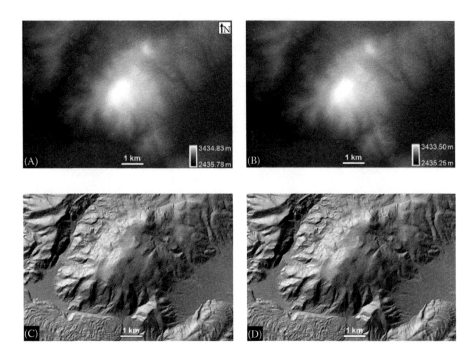

FIGURE 6.13 Original DEM and shaded DEM obtained from snow-on (A and C) and snow-off (B and D) LiDAR data collected in the upper Jerez river basin, NM (USA). Note the vertical coordinate system is NAVD88 (GEOID03) [EPSG: 5703].

Figure 6.14 shows the rasters for topographic slope, aspect, and snow-depth distribution. From Figures 6.14B through D, it can be seen that snow depths are greater on the slopes facing north, northwest, and west than those on the slopes in other directions. The contrast between thin snow depth and thick snow depth in Figure 6.14F is of interest. To further investigate the snow depth distribution, snow-on and snow-off topographic profiles are extracted from profile P–P′ (Figure 6.15). It can be seen from Figure 6.15 that west- and northwest-facing slopes have minimal snow depths, whereas east- and southeast-facing slopes have increased snow depth, especially near the hill ridge. It is believed that wind is a major causal factor for snow redistribution in this case, as also reported in Winstral and Marks (2002) and Dadic et al. (2010) for other study areas.

6.5 VOLCANIC AND IMPACT LANDFORMS

Topography plays an important role in the emplacement of lava flows (Favalli et al., 2009). High-resolution LiDAR data make it possible to investigate morphometric characteristics of lava flow, as reported in Jessop et al. (2012), Kereszturi et al. (2012), and Tarquini et al. (2012). In this section, LiDAR data for several representative volcanic landforms are presented. Since impact craters and some volcanic craters have similar shapes, LiDAR for sample impact landforms is also presented.

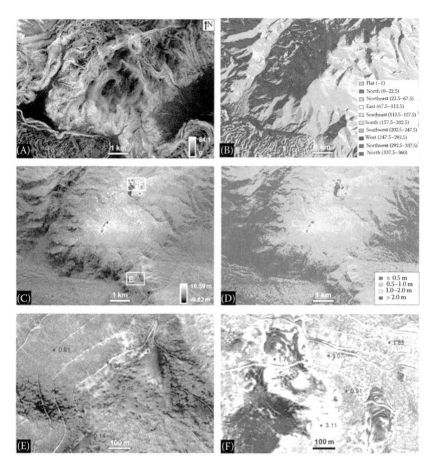

FIGURE 6.14 Topographic slope, aspect, and snow-depth distribution. (A) slope; (B) aspect; (C) snow depth map; (D) classified snow-depth map; (E) snow depth (red numbers) for box E in (C) and (F) snow depth (red text) for box F in (C).

Figure 6.16 shows LiDAR-derived DEM products in the Lunar Crater volcanic field in east-central Nevada, USA. The volcanic field includes cinder cones, maars (broad craters formed by explosive eruptions close to ground level), and basalt flows that resemble some features on the moon (Scott and Trask 1971). Of particular interest is the Lunar Crater National Natural Landmark (Figure 6.16B), a maar that is approximately 130 m deep and 1050 m wide (Figure 6.16D). As shown in Figures 6.16C and D, LiDAR data can be used to extract slope variations and cross profiles from volcanic landforms. Figure 6.17 shows selected volcanic landforms in Mauna Loa, HI, USA, and Figure 6.18 is a lava flow in Parkdale, OR, USA. Figure 6.19 is the Meteor Crater in northern Arizona, USA—the best preserved meteorite impact site on Earth. The Meteor Crater is about 1200 m in diameter and about 170 m deep, with a rim that rises about 45 m above the surrounding plains (Figures 6.19C and D). Although the profile in Figure 6.19D is similar to Figure 6.16D, the mechanisms for

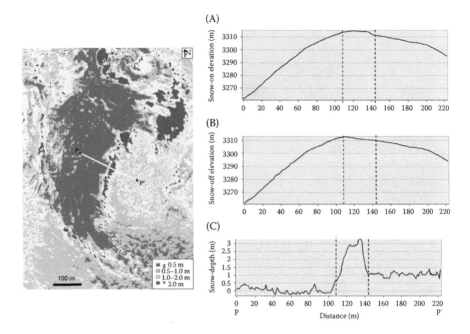

FIGURE 6.15 Classified snow depth map (left) and snow-on and snow-off topographic profiles (right). (A) Snow-on topographic profile; (B) snow-off topographic profile and (C) snow-depth profile derived from (A) and (B).

the two craters are different. The Lunar Crater in Figure 6.16 was created by explosive eruptions, whereas the Meteor Crater in Figure 6.19 was created by impact, with evidence of shock metamorphism (Kieffer 1971).

6.6 TECTONIC LANDFORMS

Numerous conceptual models of landscape evolution under tectonic and climate regimes have been proposed over the past century (Burbank and Anderson 2011). To quantify the amount of tectonic deformation, identifiable geomorphic markers are needed to provide a reference frame. These geomorphic markers include linear markers such as streams and glacial moraines, and planar markers such as terraces and alluvial fans. To calculate the rates of tectonic movement, two important parameters for geomorphic markers are needed: age and geometry. In the past several decades, new geochronologic methods have been developed for determining the age of tectonic and geomorphic markers (Burbank and Anderson 2011, Sloss et al. 2013). With increasing accuracy in dating geomorphic features, improvements in quantifying the geometry of geomorphic markers can produce more accurate rates of deformation.

The last decade has seen wide applications of LiDAR in tectonic landform studies (e.g., Arrowsmith and Zielke 2009, Hunter et al. 2009, Zielke et al. 2010, 2012, Howle et al. 2012, Dong 2014). Arrowsmith and Zielke (2009) evaluated the use

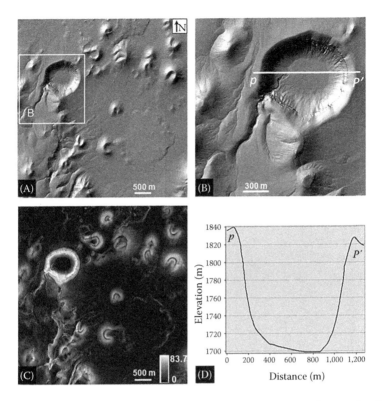

FIGURE 6.16 LiDAR-derived DEM products from part of the Lunar Crater volcanic field in east-central Nevada. (A) DEM; (B) A subarea extracted from (A); (C) Slope obtained from (A); (D) Profile extracted from *pp'* in (B). Note: The vertical coordinate system for the DEM products here is GEOID 12.

of LiDAR data for mapping recently active breaks in the Cholame segment of the south-central San Andreas Fault (SAF), and concluded that a LiDAR-only approach compares well with a combination of aerial photographic and field-based methods. In the Lake Tahoe Basin, California, USA, tectonic offsets of linear glacial moraines have been used to calculate slip rates of active normal faults obscured by dense vegetation. Howle et al. (2012) used bare-earth point cloud data to mathematically reconstruct linear lateral moraine crests on both sides of faults. The reconstructed moraine crests produced statistically significant "piercing lines" that were projected to intersection with modeled fault planes to define "piercing points" in 3D space. The results of the study yielded a two to three fold increase over previous estimates of tectonic slip rates in the Lake Tahoe region. Hunter et al. (2009) discovered a previously unmapped fault using LiDAR data near the Martis Creek Dam, Truckee, California, USA, and Székely et al. (2009) used LiDAR data in an extremely flat area, east of Neusiedlersee in Hungary, and discovered linear geomorphic features, which are several hundred meters to several kilometers long. In areas with dense vegetation cover in the U.S. Pacific Northwest and Europe, 2-m resolution DEMs derived from LiDAR data have been successfully used for delineating earthquake surface ruptures

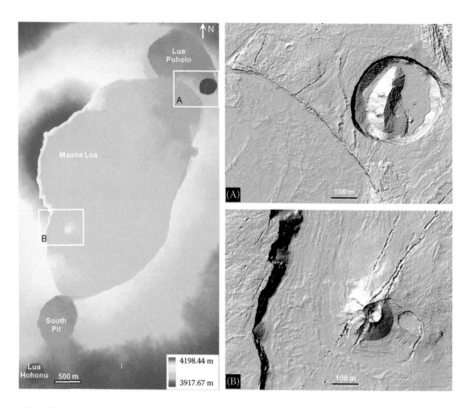

FIGURE 6.17 Selected volcanic landforms in Mauna Loa, HI (USA). Hillshaded DEMs are shown in (A) and (B). Note: The vertical coordinate system is NAVD88 (GEOID03) [EPSG: 5703].

(Haugerud et al. 2003, Sherrod et al. 2004, Cunningham et al. 2006). However, some other LiDAR studies on the northern San Andreas Fault and central Japanese mountains indicated that 2-m resolution DEMs could not identify some small tectonic breaks (Zachariasen 2008, Lin et al. 2009). Using airborne LiDAR data collected from orthogonal flight lines, Lin et al. (2013) created 0.5-m resolution DEMs along the Neodani Fault in Japan, and revealed a number of previously unknown fault scarps and active fault traces hidden under dense vegetation. Although the cost of data collection will increase with overlapping flight lines, the greater bare-earth data density, collected from different angles, will likely enhance the imaging of subtle geomorphic markers in densely forested areas.

Offset channels associated with strike-slip faults are good examples of linear geomorphic markers that can be used to determine the rate and nature of tectonic movement, and the south-central SAF in California has arguably some of the world's best-preserved tectonic landforms at 10s and 1000s of meter scale (Wallace 1975, Wallace and Schulz 1983, Wallace 1991). The most famous offset feature is the offset channel at Wallace Creek across the SAF zone in the Carrizo Plain, CA, USA (Sieh and Jahns 1984). Figure 6.20 is a hillshaded 1-m resolution DEM derived from LiDAR data that shows the offset channels in the south-central SAF. Wallace Creek is in sub-window

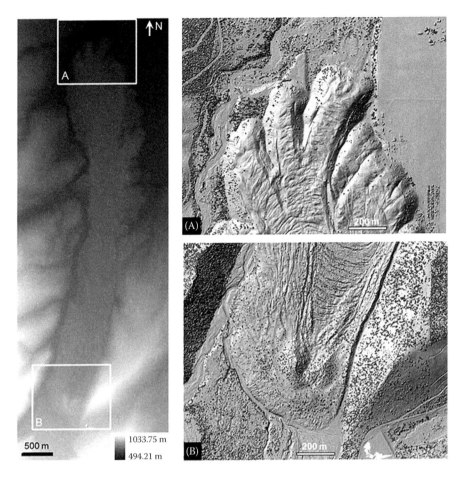

FIGURE 6.18 LiDAR data for Parkdale lava flow in Parkdale, OR (USA). Hillshaded DEMs are shown in (A) and (B). LiDAR data provided by the Oregon Department of Geology and Mineral Industries (DOGAMI) LiDAR Program.

A of Figure 6.20, and the field photo of Wallace Creek (Figure 6.21) was taken near location p_1 in sub-window A facing northwest. The distance between p_1 and p_2 is the offset distance of Wallace Creek, and a_1a_2 and b_1b_2 are two lines of topographic profiles shown in Figure 6.22. A fault scarp is visible in profile a_1a_2 (Figure 6.22A), while profile b_1b_2 (Figure 6.22B) shows a sag depression caused by lateral movement of the SAF. Beheaded channels are also visible in the central part of Figure 6.20.

Another example of offset channels is reported by Klinger and Piety (2000) in Death Valley, CA, USA, where a vertical aerial photograph was used to measure the right-lateral offset of several drainages along Furnace Creek Fault. Here a shaded DEM derived from LiDAR data is used to show the offset drainages (Figure 6.23A). The offset distances a_1a_2, b_1b_2, and c_1c_2 measured in Figure 6.23A are 309, 330, and 249 m, respectively, very close the measurements by Klinger and Piety (2000). Using

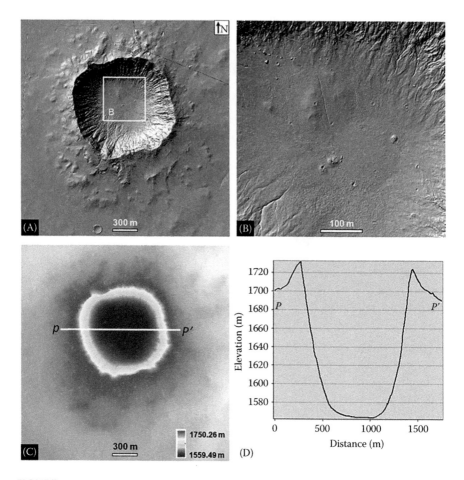

FIGURE 6.19 LiDAR-derived DEM products for Meteor Crater, AZ (USA). (A) and (B) Hillshaded DEMs; (C) DEM with colors for different elevations and (D) Topographic profile derived from P-P′ in Figure 6.19C. Note: The vertical coordinate system for DEM products is NAVD88 (GEOID 09) [EPSG: 5703].

the age of the geomorphic surface as the maximum age for the total right-lateral displacement of the drainages, Klinger and Piety (2000) calculated a slip rate of 4–9 mm/year for Furnace Creek Fault. To better reconstruct the original drainages, Figure 6.23B shows the back-slipping of the two blocks separated by the fault on hillshaded LiDAR-derived DEM.

Planar geomorphic features such as fluvial terraces, alluvial fans, and marine terraces have been widely used as geomorphic markers in tectonic deformation studies (e.g., Goy and Zazo 1986, Hetzel et al. 2002, Silva et al. 2003, Hetzel et al. 2004a, 2004b, Filocamo et al. 2009, Ramos et al. 2012, Gurrola et al. 2013, Matsu'ura et al. 2014), while erosion surfaces such as pediments (Hetzel et al. 2004a, 2004b, Hall et al. 2008) and glacis (Garcia-Tortosa et al. 2011) are less frequently used as geomorphic markers due to the difficulty in dating of such surfaces. Frankel et al. (2011)

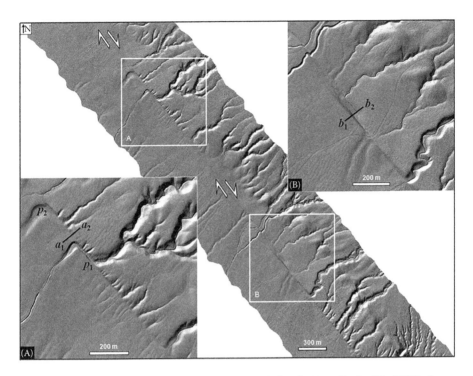

FIGURE 6.20 Offset channels in the south-central San Andreas Fault, CA (USA) shown in hillshaded DEM (1-m resolution) derived from LiDAR data. Two subareas are shown in (A) and (B).

determined new slip rates of the Death Valley–Fish Lake Valley fault system in eastern California and western Nevada by combining alluvial fan offsets measured from 1-m resolution LiDAR-derived DEM with ^{10}Be TCN and OSL ages from displaced alluvial fans. Crosby et al. (2006) extracted elevations of the marine terrace inner edges from a LiDAR-derived DEM between Fort Ross and Mendocino, CA to study deformation variations off the San Andreas Fault. Bowles and Cowgill (2012) presented a semiautomated surface classification method to identify probable marine terraces along a 70-km-long section of the northern California coast using slope and surface roughness properties obtained from LiDAR-derived DEM. In comparison with studies of linear geomorphic markers using LiDAR data, case studies of planar geomorphic markers using LiDAR data are relatively limited.

Planar geomorphic markers can help identify linear geomorphic markers in some cases. Kondo et al. (2008) provided a good example of identifying a continuous fault scarp using LiDAR data in Matsumoto, a city built on an alluvial fan in central Japan. They created a high-resolution (0.5 m) DEM after filtering out laser returns from buildings and vegetation, and identified a fault scarp of up to 2 m in height using segmented least squares fitting on the topographic profiles derived from the alluvial fan. Borehole data and archaeological studies indicate that the fault scarp is indeed in a pull-apart basin, and was formed during the most recent faulting event associated with historical earthquakes. In the Rangitaiki Plains, the fastest extending section of

FIGURE 6.21   Field photo of Wallace Creek taken near point p_1 in Figure 6.20.

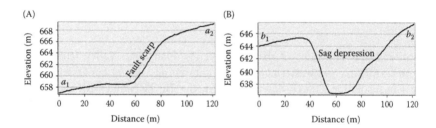

FIGURE 6.22   Topographic profiles extracted from a_1a_2 and b_1b_2 in Figure 6.20 showing a fault scarp (A) and a sag depression (B).

the onshore Taupo Rift in New Zealand, Begg and Mouslopoulou (2010) used fault-parallel and fault-normal profiles created from a LiDAR-derived DEM with 3.5 m resolution, and identified a vertical displacement of ~3 m across an active normal fault. Figure 6.24 is a slope raster created from 1-m resolution LiDAR-derived DEM of Willow Valley in Death Valley, CA. Fault scarps of a SW-NE trending fault (red line in Figure 6.24) are shown as steep slopes in Figure 6.24. Figure 6.25 is a 1-m resolution hillshaded DEM derived LiDAR data, along with locations of four profiles a_1a_2, b_1b_2, c_1c_2, and d_1d_2 shown in Figure 6.25. Uplifted terraces are revealed in profiles a_1a_2 and b_1b_2 (Figures 6.26A and B), and supported by the field study of Klinger and Piety (2000). Profile c_1c_2 (Figure 6.26C) shows the current alluvial fan, whereas profile d_1d_2 (Figure 6.26D) represents an uplifted old alluvial fan cut by streams.

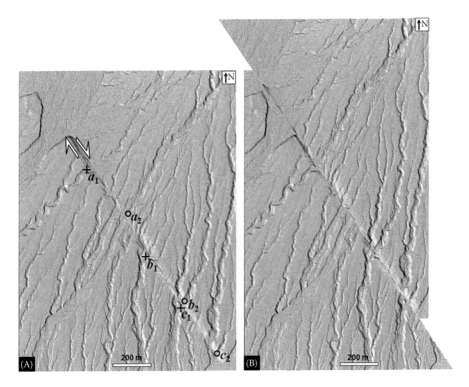

FIGURE 6.23 Offset channels along Furnace Creek Fault in Death Valley, CA (USA) shown in hillshaded DEM (1-m resolution) derived from LiDAR data. (A) Offset channels and (B) back-slipping of offset channels. a_1, b_1, and c_1 are original locations; a_2, b_2, and c_2 are shifted locations of a_1, b_1, and c_1, respectively.

6.7 LITHOLOGY AND GEOLOGIC STRUCTURES

Compared with multispectral and hyperspectral image data that has been widely used in mapping rock units and geologic structures, the application of LiDAR data in lithological and structural mapping is relatively limited, mainly due to the lack of rich spectral information of LiDAR and relatively limited availability of LiDAR data. However, the capability of LiDAR in revealing topographic details, especially in areas of dense vegetation cover, can provide unique applications in mapping rock units and geologic structures.

For lithological mapping in arid environments, integration of spectral information from optical images and texture information from radar images has proven to be effective in many studies (e.g., Dong and Leblon 2004). However in forested areas, textures from radar images usually reflect forest texture, not texture of the underlying ground surface. Grebby et al. (2010) used morphometric variables (including slope, curvature, and surface roughness) derived from a 4-m resolution LiDAR DEM to quantify the topographic characteristics of four major lithologies in the upper section of the Troodos Ophiolite, Cyprus, and produced a detailed lithological map that is more accurate than the best existing geological map in the area.

FIGURE 6.24 Slope raster created from 1-m resolution LiDAR-derived DEM of Willow Valley in Death Valley, CA (USA).

FIGURE 6.25 Hillshaded DEM created from LiDAR-derived DEM (1-m resolution) of Willow Valley in Death Valley, CA (USA).

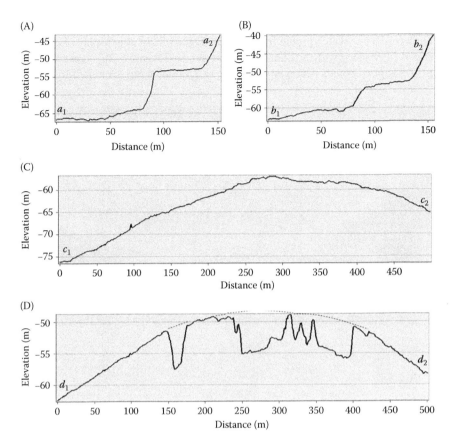

FIGURE 6.26 Topographic profiles extracted from 1-m resolution LiDAR-derived DEM along (A) a_1a_2, (B) b_1b_2, (C) c_1c_2, and (D) d_1d_2 shown in Figure 6.25.

Grebby et al. (2011) investigated the integration of airborne multispectral imagery and LiDAR-derived topographic data for lithological mapping in a vegetated section of the Troodos Ophiolite (Cyprus), and reported that LiDAR-derived topographic variables led to significant improvements of up to 22.5% in the overall mapping accuracy. Other examples of LiDAR data for rock unit mapping include Spinetti et al. (2009), Tarquini et al. (2012), and Chen et al. (2016). In addition to the previous studies on rock unit mapping, a few studies have been carried out for structural mapping using LiDAR data. Mynatt et al. (2007) used airborne LiDAR data to define the geometry of strata in Raplee Ridge in southeastern Utah, USA. Hilley et al. (2010) used airborne LiDAR data to define the geometry of exposed marker layers within the Raplee Ridge monocline in southwest Utah, USA. Le Gall et al. (2014) combined LiDAR data and echosounder data and produced a detailed structural picture of the immerged Variscan basement in the Molène archipelago, western Brittany, France.

Figure 6.27A shows a Landsat TM 8 imagery (TM7 (R), TM3 (G), TM1 (B), acquired on October 4, 2014) dragged over a vertically exaggerated Advanced Spaceborne Thermal Emission and Reflection Radiometer (ASTER) Global Digital

Elevation Map (GDEM) digital elevation model in Raplee Ridge, UT, USA, and Figure 6.27B is a sketch profile from a_1 to a_2. Figure 6.28A is a display of a LiDAR-derived DEM along with the Landsat TM 8 image, and Figure 6.28B shows a 3D view of LiDAR point clouds with a point density of 2.15 points/m², extracted from box B

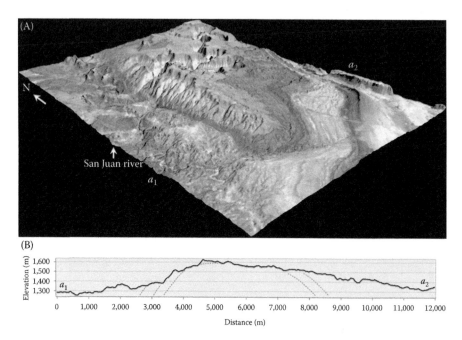

FIGURE 6.27 Raplee Ridge monocline in southwest Utah (USA). (A) Landsat TM 8 imagery [TM7 (R), TM3 (G), TM1 (B)] dragged over a vertically exaggerated (3×) ASTER GDEM digital elevation model and (B) Sketch profile from a_1 to a_2.

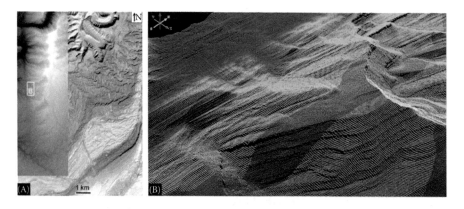

FIGURE 6.28 LiDAR data and Landsat TM 8 image for Raplee Ridge monocline in southwest Utah (USA). (A) LiDAR-derived DEM along with Landsat TM 8 image and (B) 3D view of LiDAR point clouds with a point density of 2.15 points/m².

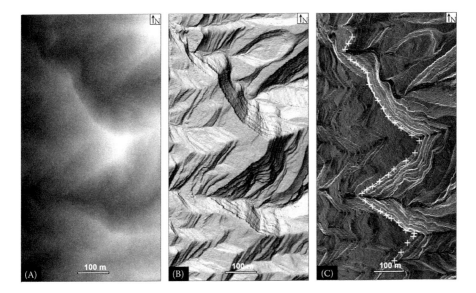

FIGURE 6.29 LiDAR-derived DEM products (1-m resolution) for box B in Figure 6.28A. (A) DEM; (B) hillshaded DEM and (C) slope raster created from DEM. White crosses in Figure 6.29C are sampling points along a rock layer (see Project 6.2 for details).

in Figure 6.28A. Figure 6.29 shows the DEM extracted from box B in Figure 6.28A, and a hillshaded DEM and slope raster. Hard rock layers are shown as bright lines in Figure 6.29C. By following the highlighted rock layers in Figure 6.29C and extracting elevations from the DEM, trend surfaces of the rock layers can be created to reconstruct the orientation of the rock layers. Project 6.2 provides a step-by-step process for constructing the trend surfaces.

PROJECT 6.1: MEASURING SAND DUNE MIGRATION USING MULTI-TEMPORAL LiDAR DATA IN WHITE SANDS DUNE FIELD, NM, USA

1. Introduction

   Understanding how sand dunes form and change has long been a research topic in Earth and planetary surface processes, yet few methods have been developed for automated detection and measurement of dune migration directions and migration rates in large dune fields. In comparison with traditional remote sensing techniques, LiDAR has provided unprecedented datasets for sand dune studies. Using the angle of repose (AOR) as a sensitive movement indicator of barchan (crescent-shaped) and transverse dunes, Dong (2015) proposed a PSTP (pairs of source and target points) method to automatically match before and after points on dune slip faces revealed by LiDAR data. The flowchart of the PSTP method is shown in Figure 6.4, and an ArcGIS add-in was created using the Python programming language to automate the whole process in Figure 6.4. The objective of this project is to use the add-in for calculating sand dune migration rates from multi-temporal LiDAR data in a study area in the WSDF, NM, USA. Before using the add-in, users will work through several steps to better understand the processes of extracting slip faces from the DEMs, and converting the centerlines of the slip faces into vector polylines.

2. Data

An area of 401 m × 802 m in WSDF is selected for this project. LiDAR data for the study area was acquired on January 24, 2009, and June 6, 2010. The LiDAR point density is about 4.19 points/m^2 (for January 24, 2009) and 4.62 points/m^2 (for June 6, 2010). The horizontal coordinate system is UTM Z13N NAD83 (CORS96) [EPSG: 26913], and the vertical coordinate system NAVD88 (Geoid 03) [EPSG: 5703]. LiDAR data acquisition and processing was completed by the National Center for Airborne Laser Mapping (NCALM). NCALM funding was provided by National Science Foundation's Division of Earth Sciences, Instrumentation and Facilities Program, EAR-1043051. Two DEM rasters in TIFF format "d20090124. tif" and "d20100606.tif" with 1 m × 1 m cell size can be downloaded (by right-clicking each file and saving it to a local folder) from the project folder at http://geography.unt.edu/~pdong/LiDAR/Chapter6/Project6.1/.

3. Project Steps

1. Open an empty Word document so that you can copy any results from the following steps to the document. To copy the whole screen to your Word document, press the PrtSc (print screen) key on your keyboard, then open your Word document and click the "Paste" button or press Ctrl+V to paste the content into your document. To copy an active window to your Word document, press Alt+PrtSc, then paste the content into your document.

2. Open ArcMap, load the Spatial Analyst Extension, 3D Analyst Extension, and ArcScan, and then add "d20090124.tif" and "d20100606. tif" to ArcMap (Figure 6.30 and Figure 6.31).

3. Create DEM profile. To better understand the concepts of slip face and angle of repose for sand dunes, open the Customize menu of ArcMap and select Toolbars → 3D Analyst to open the 3D Analyst toolbar. Select d20090124.tif as the 3D Analyst Layer on the toolbar, click the Interpolate Line tool, draw a straight line from left to right in the central part of the DEM for January 24, 2009, and click the Profile Graph icon on the 3D Analyst Toolbar to display the DEM profile (see Figure 6.32 for a sample).

4. Create empty polyline shapefile in ArcCatalog, which will be used for storing polylines for slip face centers later in the project. Open ArcCatalog and select the project folder for Project 6.1, then right click the ArcCatalog window and select "New" → "Shapefile..." to create a polyline shapefile "polyline20090124.shp". To define the spatial

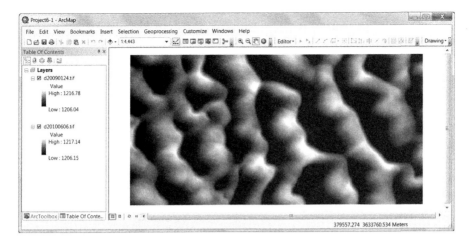

FIGURE 6.30   LiDAR-derived DEM (1-m resolution) of January 24, 2009, for the 401 m × 802 m test area.

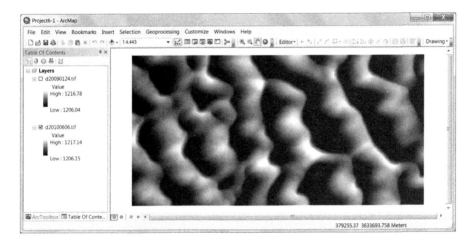

FIGURE 6.31   LiDAR-derived DEM (1-m resolution) of June 6, 2010, for the 401 m × 802 m test area.

reference of polyline20090124.shp, click the "Edit" button and import the spatial reference properties of the DEM raster "d20090124.tif". Add polyline20090124.shp to ArcMap.

5. Create a slope raster from the 2009 DEM. Open ArcToolbox → Spatial Analyst Tools →Surface → Slope, use "d20090124.tif" as the input raster to create the output slope raster "slp20090124" (Figure 6.33). The maximum slope is 35.604° in the slope raster. However, if raster cells with a slope value over 34.8° are selected using the Raster Calculator, it can be seen that only several isolated cells are selected. These isolated cells with

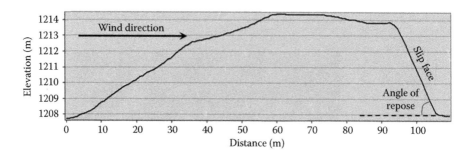

FIGURE 6.32 Sample profile for a single dune in the 2009 DEM. Note the vertical exaggeration.

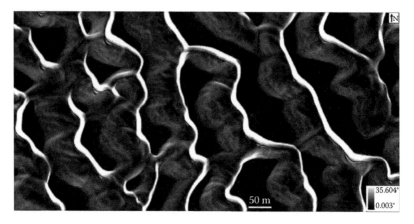

FIGURE 6.33 Slope raster derived from the DEM for January 24, 2009.

slope values greater than 34.8° are probably caused by bushes or other objects in the desert, and can be easily removed in the next steps.

6. Extract slip faces from the slope raster. In this project, slip faces are identified by selecting slopes greater than 30° and less than 35°. Open ArcToolbox → Spatial analyst Tools → Map Algebra → Raster Calculator, use Con((("slp20090124" > 30) and ("slp20090124" < 35)), 1, 0) as the expression to create binary raster "face20090124" where 1's are for slip faces (black cells in Figure 6.34), and 0's for other cells. Note: You should have full control over the output folder of the binary raster, otherwise you may not be able to edit the binary raster in Step 7 below.

7. Edit binary raster for slip faces. Open the "Customize" menu of ArcMap and select Toolbars → Editor. Right-click the binary raster layer "face20090124" in the ArcMap table of contents, then select Editing Features → Start Editing. Then open the Customize menu in ArcMap and select Toolbars → ArcScan. Select "face20090124" as the ArcScan Raster Layer for the ArcScan Toolbar (Figure 6.35).

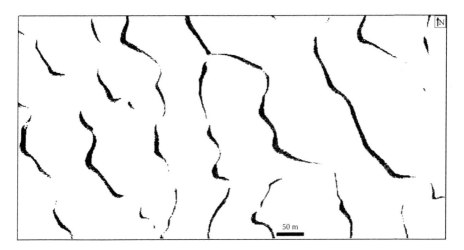

FIGURE 6.34   Binary raster "face20090124" for slip faces.

FIGURE 6.35   ArcScan Toolbar in ArcMap.

Open the Raster Cleanup menu on the ArcScan Toolbar (Figure 6.35) and select "Start Cleanup" to enable the tools under the Raster Cleanup menu. Select the mathematical morphological operation "Closing..." under the Raster Cleanup menu, put 1 as the number of pixels, and click OK remove noises in the binary raster.

8. Create polylines from slip face centerlines through vectorization. Open the Vectorization menu on the ArcScan Toolbar (Figure 6.35) and use the default settings and options, then click "Show Preview" under the Vectorization menu to see the preview. In this project, the default settings work fine, so you do not need to make any changes. Right-click the empty polyline shapefile "polyline20090124.shp" on the ArcMap table of contents, and select Editing Features → Start Editing. You will be asked if you want to save the edits to the raster "face20090124". Click "Yes" to save the edits. Now select "Generate Features..." under the Vectorization menu of the ArcScan Toolbar, use "polyline20090124" as the template, and click OK to generate polyline features for slip face centerlines derived from the 2009 DEM. On the Editor toolbar, select Edit → Stopping Editing to save the edits to slip face centerlines. Figure 6.36 shows the slip face centerlines (polyline20090124.shp) over the 2009 DEM (d20090124.tif).

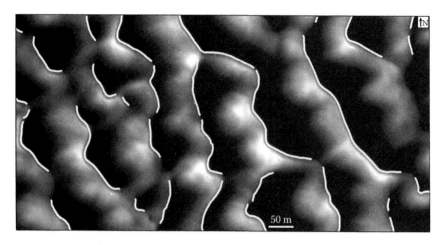

FIGURE 6.36 Extracted slip face centerlines (white) over 2009 DEM.

As can be seen from Steps 4–8, it would be time-consuming to repeat the process using the DEM for June 6, 2010 (d20100606.tif) and follow the flowchart in Figure 6.4 to obtain the final results. Therefore, an ArcGIS add-in has been created by the author using the Python programming language to automate the whole process. The ArcGIS add-in "DuneMigration.esriaddin" can be downloaded from the project folder at http://geography.unt.edu/~pdong/LiDAR/Chapter6/Project6.1/. In the following steps, the add-in will be installed, and source directions and dune migration rates will be calculated using the add-in toolbar.

9. Install the "Dune Migration" add-in. Double click the file "DuneMigration.esriaddin" in the project folder to open the add-in Installation Utility window (Figure 6.37). Click the "Install add-in" button, and you should see a pop-up message: "Installation Succeeded."

10. Load the Dune Migration add-in toolbar. Open the "Customize…" menu in ArcMap and select "Add-in Manager…". Then select the "DuneMigration" add-in and click "Customize…" (Figure 6.38). In the Customize window, check the Dune Migration toolbar, and then click "Close". The Dune Migration toolbar should appear (Figure 6.39).

11. The parameters for the Dune Migration Toolbar in Figure 6.39 are explained below, and the results are shown in Figures 6.40 through 6.42.

    a. DEM1: The first DEM raster which can be created from LiDAR data or other data sources. The data acquisition date is contained in the DEM layer name in the format of YYYYMMDD, and the YYYYMMDD string can be any where in the DEM name as long as it is the first eight numbers; for example, A20090124DEM1. The DEM layer name can be changed by the user in ArcMap, and can be different from the actual file name.

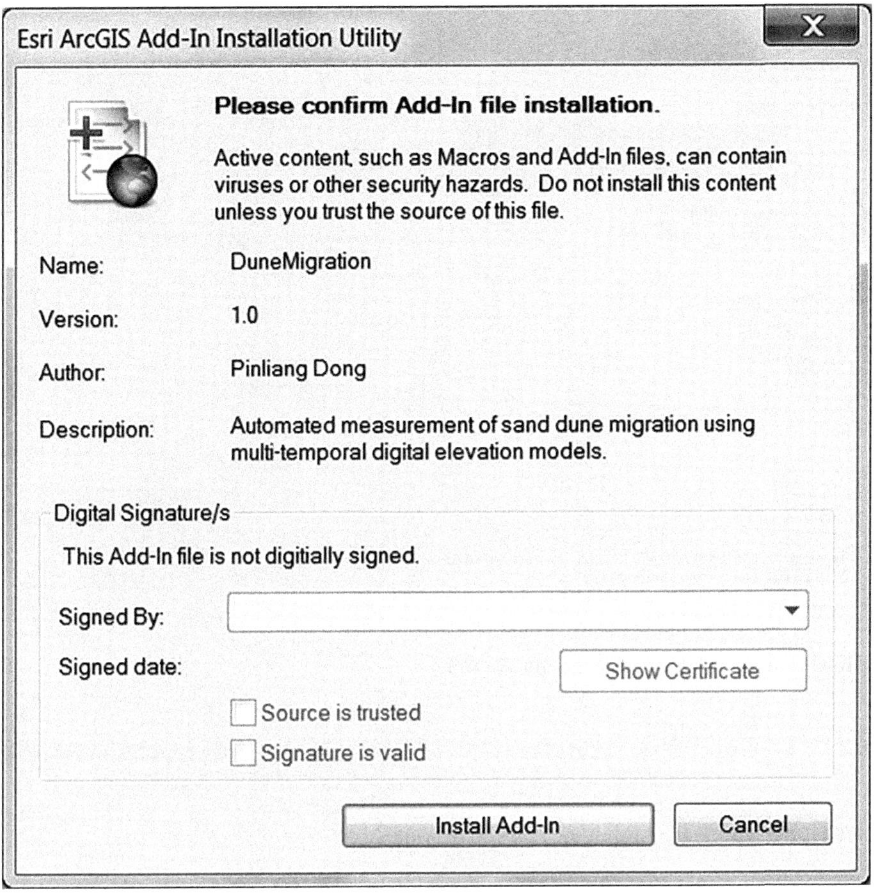

FIGURE 6.37   ArcGIS add-in installation utility.

  b. DEM2: The second DEM raster (similar to DEM1). The dates for
     DEM1 and DEM2 are used for calculating the time interval (num-
     ber of days) between DEM1 and DEM2, which will be used to con-
     vert dune migration distance into migration rate at each sampling
     point.
  c. AOR: Angle of repose for sand dune slip faces. AOR is usually
     around 34°, depending on the sand grain size, shape, and mois-
     ture content. Users can select/input a range, such as 30–35, as AOR
     values.
  d. Min-Dist: The minimum distance between two random points. The
     unit of distance is the same as the linear unit of the DEM layers.
  e. Radius: The search radius used to identify the nearest source point
     around a random target point. The unit of radius is the same as the
     linear unit of the DEM layers.

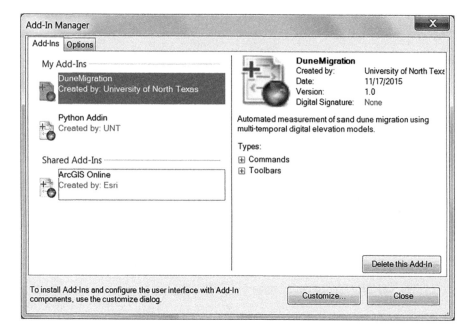

FIGURE 6.38   Add-in Manager for ArcMap.

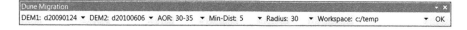

FIGURE 6.39   Dune Migration Toolbar for ArcMap.

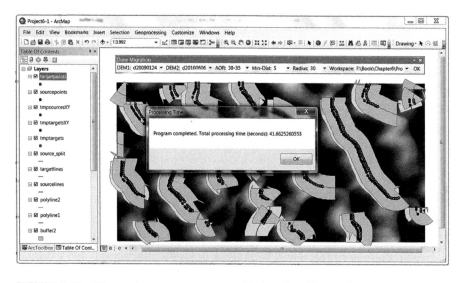

FIGURE 6.40   The test datasets were processed in less than 42 seconds.

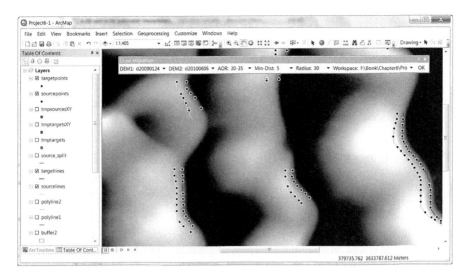

FIGURE 6.41 Target points on target lines (red, for June 6, 2010), and source points for source lines (Green, for January 24, 2009).

FID	Shape *	NEAR_DIST	NEAR_X	NEAR_Y	NEAR_ANGLE	Azimuth0	Azimuth1	Diff_Azi	m_Rate
0	Point	10.045656	379426.9799	3634010.15506	-174.289407	264.289	174.289	90	7.36278
1	Point	9.590955	379427.487759	3634005.07647	-174.289407	264.289	174.289	90	7.02952
2	Point	10.108606	379429.487476	3633996.87115	-155.556045	245.556	155.556	90	7.40892
3	Point	10.168364	379431.583568	3633992.25974	-155.556045	245.556	155.556	90	7.45272
4	Point	10	380049.975701	3633979.95514	180	270	0	270	7.32932
5	Point	11.455775	380049.975701	3633975.2855	180	270	0	270	8.3963
6	Point	12.587863	380051.840124	3633966.60378	-161.565051	251.565	161.565	90	9.22604
7	Point	9	379605.975701	3634021.16765	180	270	0	270	6.59639
8	Point	9	379605.975701	3634015.97633	180	270	0	270	6.59639
9	Point	7.111591	379608.801112	3634007.25258	-150.255119	240.255	150.255	90	5.21231
10	Point	7.025692	379610.143614	3634004.5254	-165.963757	255.964	165.964	90	5.14935
11	Point	7.781547	379614.339523	3633997.31572	-139.085617	229.086	139.086	90	5.70334
12	Point	7.65866	379617.628123	3633993.52118	-139.085617	229.086	139.086	90	5.61328
13	Point	6.019416	379620.667891	3633990.01376	-139.085617	229.086	139.086	90	4.41182
14	Point	6.37817	379623.957318	3633986.21826	-139.085617	229.086	139.086	90	4.67476
15	Point	7.602758	379629.069294	3633977.41599	-140.710593	230.711	140.711	90	5.5723
16	Point	7.08949	379632.384508	3633973.36407	-140.710593	230.711	140.711	90	5.19611
17	Point	7.972901	379638.737329	3633964.16674	-136.974934	226.975	136.975	90	5.84359
18	Point	6.66975	379642.188065	3633960.46952	-136.974934	226.975	136.975	90	4.88847
19	Point	5.233663	379645.504543	3633956.91615	-136.974934	226.975	136.975	90	3.83592

FIGURE 6.42 Attributes of target points. NEAR_DIST—migration distance, Azimuth0—source direction, and m_Rate—migration rate (m/year).

    f. Workspace: The folder for output rasters and shapefile. To ensure the geoprocessing steps are not affected by any existing files, there should be no existing files or folders in the workspace before a users clicks the OK button; otherwise, a warning message will pop up.

g. OK: Click OK to run the program. If there are any errors in the parameters on the toolbar, error messages will pop up. Results from the test data are shown in the following figures.

12. Save your ArcMap project.

13. Questions: (a) How can you show histograms of source directions and migration rates? (b) How can you create a continuous raster in ArcGIS to show dune migration rates in the study area?

PROJECT 6.2: DERIVING TREND SURFACES OF SIMPLE FOLDS USING LiDAR DATA IN RAPLEE RIDGE, UT, USA

1. Introduction

Geologists use two different compass bearings, strike and dip, to define the orientation of rock strata in 3D space. The intersection between a dipping rock layer and an imaginary horizontal place is a line. Strike is the compass bearing (relative to north) of the intersection line, while dip is the direction of maximum inclination down from strike, and is always perpendicular to strike. An angular measurement, dip magnitude, is the smaller of two angles formed by the intersection of an imaginary horizontal plane and a dipping rock layer. The 3D orientation of rock layers and simple folds can be represented by trend surfaces. A trend surface is a smooth surface defined by a mathematical function (a polynomial) that fits the input sample points using the least-squares fitting. The first-order, second-order, and third-order trend surfaces are defined by the following equations:

First-order:

$$f(x,y) = c_0 + c_1 x + c_2 y \tag{6.1}$$

Second-order:

$$f(x,y) = c_0 + c_1 x + c_2 y + c_3 x^2 + c_4 xy + c_5 y^2 \tag{6.2}$$

Third-order:

$$f(x,y) = c_0 + c_1 x + c_2 y + c_3 x^2 + c_4 xy + c_5 y^2 + c_6 x^3 + c_7 x^2 y + c_8 xy^2 + c_9 y^3 \tag{6.3}$$

where x and y are the (x, y) coordinates of input points, $c_0 \sim c_9$ are coefficients obtained by solving a set of simultaneous linear equations, and $f(x, y)$ is the output z value at (x, y). A first-order trend surface is a flat or tilted plane without any bending; a second-order trend surface is a concave or convex surface, and a third-order trend surface has two bends. For simple rock layers and folds, the above equations should be enough to capture general trends of the surfaces.

The basic idea of this project is to create a point shapefile and add points to the shapefile along the outcrops of a rock layer, then use the points to extract elevations (z values) from the DEM. Finally, a trend surface representing the orientation of the rock layer can be created from the (x, y, z) points.

2. Data

In this project, LiDAR data collected from Raplee Ridge, UT, USA on February 24, 2005 is used to derive trend surfaces of rock layers and simple folds. LiDAR data acquisition and processing was completed by the NCALM. NCALM funding was provided by National Science Foundation's Division of Earth Sciences, Instrumentation and Facilities Program. EAR-1043051. The LiDAR point density is about 2.15 points/m^2. The horizontal coordinate system is UTM z12 N NAD83 (CORS96) [EPSG: 26912], vertical coordinate system NAVD88 (Geoid 03) [EPSG: 5703]. A DEM raster in TIFF format "dem.tif" with 1 m × 1 m cell size can be downloaded (by right-clicking each file and saving it to a local folder) from the project folder at http://geography.unt.edu/~pdong/LiDAR/Chapter6/Project6.2/.

3. Project Steps

1. Open an empty Word document so that you can copy any results from the following steps to the document. To copy the whole screen to your Word document, press the PrtSc (print screen) key on your keyboard, then open your Word document and click the "Paste" button or press Ctrl+V to paste the content into your document. To copy an active window to your Word document, press Alt+PrtSc, then paste the content into your document.

2. Add DEM data. Open ArcMap, turn on the Spatial Analyst extension, and then add raster "dem.tif" from the project folder (Figure 6.43).

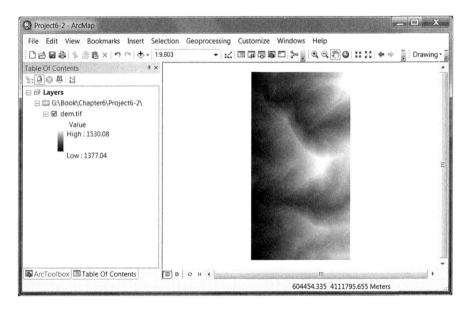

FIGURE 6.43 LiDAR-derived DEM (1 m × 1 m cell size) of a study area in Raplee Ridge, UT, USA.

3. Create hillshaded DEM. Open ArcToolbox → Spatial Analyst Tools → Surface → Hillshade. Use the hillshade tool (Figure 6.44) to create a hillshaded DEM (Figure 6.45) to help interpret topographic features of the study area.

4. Create slope raster. Open ArcToolbox → Spatial Analyst Tools → Surface → Slope, and use dem.tif as the input raster to a slope "slope"

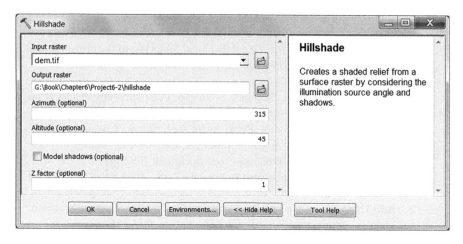

FIGURE 6.44   Hillshade tool in ArcGIS.

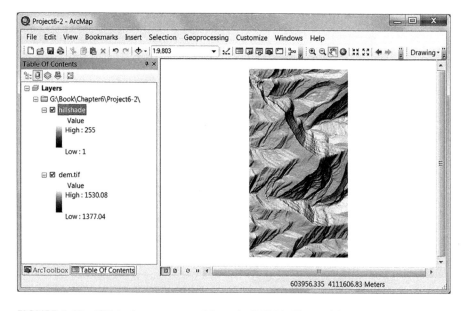

FIGURE 6.45   Hillshade raster created from the DEM in Figure 6.20.

(Figure 6.46). Due to the differences in the resistance of the rock layers to the weathering processes, outcrops of some rock layers may show relatively steep slopes (bright tones in Figure 6.46).

5. Create new shapefile for sample points. Open ArcCatalog and select the project folder for Project 6.2, then right click the ArcCatalog window and select "New" → "Shapefile…" to create a point shapefile "samples.shp". To define the spatial reference of samples.shp, click the "Edit" button and import the spatial reference properties of the DEM raster "dem.tif".

6. Add sample points to shapefile. Add the empty point shapefile "samples.shp" to ArcMap and change the symbol of the shapefile to red cross or any other point symbol. Load the Editor toolbar, and select "Start Editing" in the dropdown menu "Editor", then select "Editing Windows" → "Create Features". In the "Create Features" window on the right side of ArcMap, click "samples", and then click "Point" in the "Construction Tools" window on the lower-right corner of ArcMap (Figure 6.47). Now you can zoom in to a rock layer (a bright feature in the slope raster), and start adding sample points along the feature. If a high-resolution remotely sensed image such as IKONOS or GeoEye is available and co-registered to the LiDAR-derived DEM, the image can be added to ArcMap and used as reference when adding the sample points. To complete the editing process, select "Stop Editing" in the dropdown menu "Editor", and save the edits.

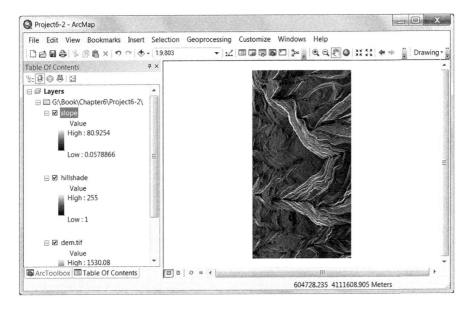

FIGURE 6.46 Slope raster created from the DEM in Figure 6.43.

FIGURE 6.47 Adding sample points along a rock layer based on the slope raster.

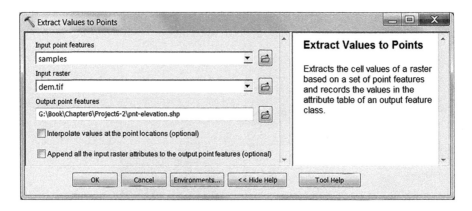

FIGURE 6.48 Extracting elevation values to points.

7. Extract elevations using sample points. The point shapefile "samples.shp" created in Step 6 can be used to extract elevations at individual sample locations from the DEM. Open ArcToolbox → Spatial Analyst Tools → Extraction → Extract Values to Points, and extract z values to a new point shapefile "pnt-elevations.shp" where z values are saved in the RASTERVALU field (Figure 6.48).

8. Create trend surfaces. Open ArcToolbox → Spatial Analyst Tools → Interpolation → Trend, and set the parameters as in Figure 6.49 to create a second-order trend surface from sample points in "pnt-elevation.shp". Similarly, first-order and third-order trend surfaces can also be created

using the Trend tool. Table 6.1 lists the coefficients, RMS errors, and Chi-square values for the first-order, second-order, and third-order trend surfaces created from the 35 points in Figure 6.29C. Since the rock layer used in this project is part of an anticline, the first-order trend surface may not be the best option for representing the orientation of the rock layer, as indicated in the RMS error and Chi-square values in Table 6.1.

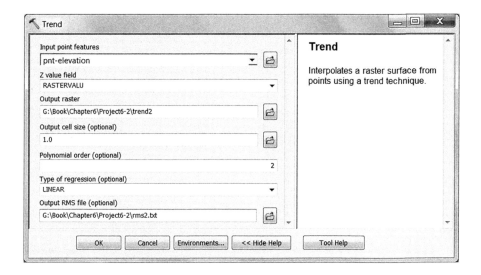

FIGURE 6.49 Generating a second-order trend surface from sample points.

TABLE 6.1
Coefficients, RMS Errors, and Chi-Square Values of Three Trend Surfaces

Order of Polynomial	First-Order	Second-Order	Third-Order
Coefficients	$c_0 = -559588.0989$	$c_0 = -330290248.7063$	$c_0 = -141724654.4038$
	$c_1 = 0.5203$	$c_1 = 871.4967$	$c_1 = 170.9755$
	$c_2 = 0.0599$	$c_2 = 32.4302$	$c_2 = -2.1991$
		$c_3 = -0.0009$	$c_3 = 0.0003$
		$c_4 = 4.7504e{-}005$	$c_4 = 4.4767e{-}005$
		$c_5 = -7.4263e{-}006$	$c_5 = 1.1979e{-}006$
			$c_6 = -6.2593e{-}010$
			$c_7 = -8.2176e{-}012$
			$c_8 = 1.5412e{-}012$
			$c_9 = -7.7478e{-}013$
RMS error	5.1103	3.1367	3.1367
Chi-square	914.04785	344.3583	344.3589

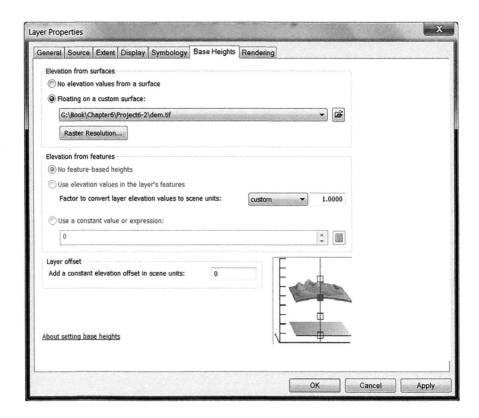

FIGURE 6.50   Setting base heights for a raster scene layer.

9. 3D visualization. To create 3D visualization of the DEM, sample points, and trend surfaces, open ArcScene and add the DEM raster "dem.tif", the second-order trend surface raster "trend2", and the sample points "pnt-elevation.shp" as scene layers. You can change the background color of the scene, and the symbology or transparency of each scene layer as desired. It is important to set the base height property for each layer to obtain a 3D view. For raster scene layers, open the Layer Properties form and select the Base Heights tab, select "Floating on a custom surface", then select the corresponding raster for the raster scene layer and click "Apply" (Figure 6.50). You can also change raster resolution (Figure 6.50). For vector scene layers, select "Use a constant value or expression" in the Base Heights tab (Figure 6.51), and click the expression builder button to select a field for z values (field [RASTERVALU] in this case). The DEM, sample points, and second-order trend surface are shown in ArcScene in Figure 6.52.

10. Save your ArcMap and ArcScene projects and Word document.

11. Question: (a) Suppose the second-order trend surface created in Step 8 can be used for representing the orientation of the rock layer, how can you calculate the dip direction, dip magnitude, and strike at any location

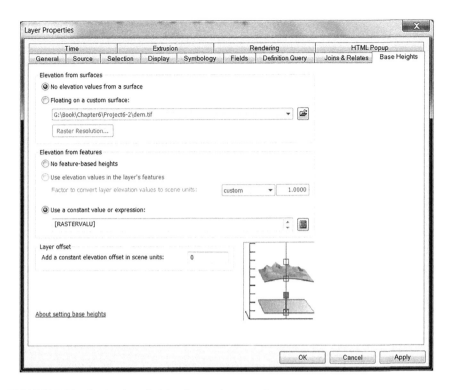

FIGURE 6.51   Setting base heights for a point scene layer.

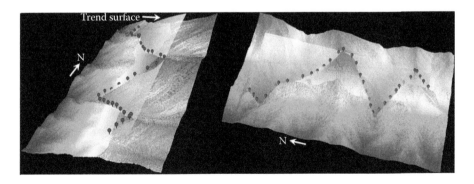

FIGURE 6.52   A mosaic of ArcScene visualization of DEM, sample points, and second-order trend surface.

of the rock layer using ArcGIS? (b) If two parallel rock layers are represented by two first-order trend surfaces L_1 and L_2 (Figure 6.53), how can you calculate the distance T between the two layers in ArcGIS? Note: If L_1 is the top and L_2 is the bottom of a rock layer, T is the true thickness of the rock layer. (Hint: Extract z values (elevations) z_1 and z_2 from the two surfaces using a single point, and get Δz as the first step.)

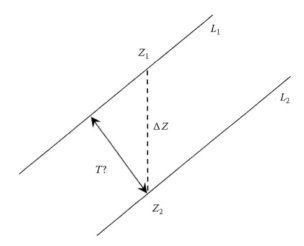

FIGURE 6.53 Two first-order trend surfaces L_1 and L_2 for two parallel rock layers.

REFERENCES

Alhajraf, S., 2004. Computational fluid dynamic modeling of drifting particles at porous fences. *Environmental Modelling and Software*, 19: 163–170.

Araújo, A.D., Parteli, E.J.R., Pöschel, T., Andrade, J.S., and Herrmann, H.J., 2013. Numerical modeling of the wind flow over a transverse dune. *Scientific Reports*, 3: 2858, doi: 10.1038/srep02858.

Arrowsmith, J.R., and Zielke, O., 2009. Tectonic geomorphology of the San Andreas Fault zone from high resolution topography: An example from the Cholame segment. *Geomorphology*, 113: 70–81.

Bagnold, R.A., 1941. *The Physics of Blown Sand and Desert Dunes*, Chapman and Hall, London, 265 p.

Bailey, S.D., and Bristow, C.S., 2004. Migration of parabolic dunes at Abberffraw, Anglesey, North Wales. *Geomorphology*, 59: 165–174.

Baitis, E., Kocurek, G., Smith, V., Mohrig, D., Ewing, R.C., and Peyret, A.-P.B., 2014. Definition and origin of the dune-field pattern at White Sands, New Mexico. *Aeolian Research*, 15: 269–287.

Barrio-Parra, F., and Rodríguez-Santalla, I., 2014. A free cellular model of dune dynamics: Application to El Fangar spit dune system (Ebro Delta, Spain). *Computers and Geosciences*, 62: 187–197.

Begg, J.G., and Mouslopoulou, V., 2010. Analysis of late Holocene faulting within an active rift using lidar, Taupo Rift, New Zealand. *Journal of Volcanology and Geothermal Research*, 190: 152–167.

Bourke, M.C., Lancaster, N., Fenton, L.K., Parteli, E.J.R., Zimbelman, J.R., and Radebaugh, J., 2010. Extraterrestrial dunes: An introduction to the special issue on planetary dune systems. *Geomorphology* 121: 1–14.

Bowles, C.J., and Cowgill, E., 2012. Discovering marine terraces using airborne LiDAR along the Mendocino-Sonoma coast, northern California. *Geosphere*, 8: 386–402.

Bridges, N.T., Ayoub, F., Avouac, J.-P., Leprince, S., Lucas, A., and Mattson, S., 2012. Earth-like sand fluxes on Mars. *Nature*, 485: 339–342, doi: 10.1038/nature11022.

Bull, J.M., Miller, H., Gravley, D.M., Costello, D., Hikuroa, D.C.H., and Dix, J.K., 2010. Assessing debris flows using LIDAR differencing: 18 May 2005 Matata event, New Zealand. *Geomorphology*, 124: 75–84.

Burbank, D.W., and Anderson, R.S., 2011. *Tectonic Geomorphology*, Wiley-Blackwell, New York, 472 p.

Cavalli, M., Tarolli, P., Marchi, L., and Fontana, G.D., 2008. The effectiveness of airborne LiDAR data in the recognition of channel-bed morphology. *CATENA*, 73: 249–260.

Crosby, C.J., Arrowsmith, J.R., and Prentice, C.S., 2006. Application of LiDAR data to constraining a late Pleistocene slip rate and vertical deformation of the Northern San Andreas Fault, Fort Ross to Mendocino, California: Collaborative research between Arizona State University and the U.S. Geological Survey. In: *3rd Annual Northern California Earthquake Hazards Workshop Abstract Volume*, Menlo Park, CA, 18–19 January, 2006.

Cunningham, D., Grebby, S., Tansey, K., Gosar, A., and Kastelic, V., 2006. Application of airborne LiDAR to mapping seismogenic faults in forested mountainous terrain, southeastern Alps, Slovenia. *Geophysical Research Letters*, 33: L20308.

Dadic, R., Mott, R., Lehning, M., and Burlando, P., 2010. Wind influence on snow depth distribution and accumulation over glaciers. *Journal of Geophysical Research*, 115: F01012.

Davis, W.M., 1905. The geographical cycle in an arid climate. *The Journal of Geology*, 13: 381–407.

Dong, P., 2012. Editorial: Applications of light detection and ranging (LiDAR) in geosciences. *Journal of Geology and Geosciences*, 1:e102, doi: 10.4172/jgg.1000e102.

Dong, P., 2014. LiDAR data for characterizing linear and planar geomorphic markers in tectonic geomorphology. *Journal of Geophysics and Remote Sensing*, 4: 136. doi: 10.4172/2169-0049.1000136.

Dong, P., 2015. Automated measurement of sand dune migration using multi-temporal LiDAR data and GIS. *International Journal of Remote Sensing*, 36: 5526–5547.

Dong, P., and Guo, H.D., 2012. A framework for automated assessment of post-earthquake building damage using geospatial data. *International Journal of Remote Sensing*, 33: 81–100.

Dong, P., and Leblon, B., 2004. Rock unit discrimination on Landsat-TM, SIR-C and RADARSAT images using spectral and textural information. *International Journal of Remote Sensing*, 25: 3745–3768.

Dong, Z., Wang, T., and Wang, X., 2004. Geomorphology of the megadunes in the Badain Jaran Desert. *Geomorphology*, 60: 191–203.

Dong, Z., Wang, X., and Chen, G., 2000. Monitoring sand dune advance in the Taklimakan Desert. *Geomorphology*, 35: 219–231.

Dye, D.G., and Tucker, C.J., 2003. Seasonality and trends of snow-cover, vegetation index, and temperature in northern Eurasia. *Geophysical Research Letters*, 30: 1405.

Easterbrook, D.J., 1999. *Surface Professes and Landforms*, Prentice Hall, New York, 546 p.

Elbelrhiti, H., Claudin, P., and Andreotti, B., 2005. Field evidence for surface-wave-induced instability of sand dunes. *Nature*, 437: 720–723.

Ewing, R.C., and Kocurek, G., 2010a. Aeolian dune-field pattern boundary conditions. *Geomorphology*, 114: 175–187.

Ewing, R.C., and Kocurek, G.A., 2010b. Aeolian dune interactions and dune-field pattern formation: White Sands Dune Field, New Mexico. *Sedimentology*, 57: 1199–1219.

Ewing, R.C., McDonald, G.D., and Hayes, A.G., 2015. Multi-spatial analysis of aeolian dune-field patterns. *Geomorphology*, 240: 44–53.

Favalli, M., Mazzarini, F., Pareschi, M.T., and Boschi, E., 2009. Topographic control on lava flow paths at Mount Etna, Italy: Implications for hazard assessment. *Journal of Geophysical Research*, 114: F01019.

Fenton, L.K., 2006. Dune migration and slip face advancement in the Rabe Crater dune field, Mars. *Geophysical Research Letters*, 33: L20201, doi: 10.1029/2006GL027133.

Fewtrell, T.J., Duncan, A., Sampson, C.C., Neal, J.C., and Bates, P.D., 2011. Benchmarking urban flood models of varying complexity and scale using high resolution terrestrial LiDAR data. *Physics and Chemistry of the Earth*, Parts A/B/C 36: 281–291.

Filocamo, F., Romano, P., Di Donato, V., Esposito, P., Mattei, M., Porreca, M., Robustelli, G., and Russo Ermolli, E., 2009. Geomorphology and tectonics of uplifted coasts: New chronostratigraphical constraints for the Quaternary evolution of Tyrrhenian North Calabria (southern Italy). *Geomorphology*, 105: 334–354.

Frankel, K.L., Dolan, J.F., Owen, L.A., Ganev, P., and Finkel, R.C., 2011. Spatial and temporal constancy of seismic strain release along an evolving segment of the Pacific–North America plate boundary. *Earth and Planetary Science Letters*, 304: 565–576.

García-Tortosa, F.J., Alfaro, P., Sanz de Galdeano, C., and Galindo-Zaldívar, J., 2011. Glacis geometry as a geomorphic marker of recent tectonics: The Guadix–Baza basin (South Spain). *Geomorphology*, 125: 517–529.

Gay, S.P., 1999. Observations regarding the movement of barchan sand dunes in the Nazca to Tanaca area of southern Peru. *Geomorphology*, 27: 279–293.

Gigli, G., and Casagli, N., 2011. Semi-automatic extraction of rock mass structural data from high resolution LIDAR point clouds. *International Journal of Rock Mechanics and Mining Sciences*, 48: 187–198.

Glenn, N.F., Streutker, D.R., Chadwick, D.J., Thackray, G.D., and Dorsch, S.J., 2006. Analysis of LiDAR-derived topographic information for characterizing and differentiating landslide morphology and activity. *Geomorphology*, 73: 131–148.

Goy, J.L., and Zazo, C., 1986. Synthesis of the Quaternary in the Almería littoral. Neotectonic activity and its morphologic features, Western Betics, Spain. *Tectonophysics*, 130: 259–270.

Grebby, S., Cunningham, D., Naden, J., and Tansey, K., 2010. Lithological mapping of the Troodos ophiolite, Cyprus, using airborne LiDAR topographic data. *Remote Sensing of Environment*, 114: 713–724.

Grebby, S., Naden, J., Cunningham, D., and Tansey, K., 2011. Integrating airborne multispectral imagery and airborne LiDAR data for enhanced lithological mapping in vegetated terrain. *Remote Sensing of Environment*, 115: 214–226.

Grohmann, C.H., and Sawakuchi, A.O., 2013. Influence of cell size on volume calculation using digital terrain models: A case of coastal dune fields. *Geomorphology*, 180–181: 130–136.

Gurrola, L.D., Keller, E.A., Chen, J.H., Owen, L.A., and Spencer, J.Q., 2013. Tectonic geomorphology of marine terraces: Santa Barbara fold belt, California. *Geological Society of America Bulletin*, 126: 219–233.

Ha, S., Dong, G., and Wang, G., 1999. Morphodynamic study of reticulate dunes at southeastern fringe of the Tengger Desert. *Science China*, 42: 207–215.

Hall, S.R., Farber, D.L., Audin, L., Finkel, R.C., and Mériaux, S., 2008. Geochronology of pediment surfaces in southern Peru: Implications for quaternary deformation of the Andean forearc. *Tectonophysics*, 459: 186–205.

Haugerud, R.A., Harding, D.J., Johnson, S.Y., Harless, J.L., Weaver, C.S., and Sherrod, B.L., 2003. High-resolution topography of the Puget Lowland, Washington—A bonanza for earth science. *GSA Today*, 13: 4–10.

Hersen, P., 2004. On the crescentic shape of barchans dunes. *The European Physical Journal B*, 37: 507–514.

Hetzel, R., Niedermann, S., Tao, M.X., Kubik, P.W., Ivy-Ochs, S., Gao, B., and Strecker, M.R., 2002. Low slip rates and long-term preservation of geomorphic features in Central Asia. *Nature*, 417: 428–432.

Hetzel, R., Tao, M.X., Stokes, S., Niedermann, S., Ivy-Ochs, S., Gao, B., and Strecker, M.R., 2004a. Implications of the fault scaling law for the growth of topography: Mountain ranges in the broken foreland of NE Tibet. *Terra Nova*, 16: 157–162.

Hetzel, R., Tao, M.X., Stokes, S., Niedermann, S., Ivy-Ochs, S., Gao, B., Strecker, M.R., and Kubik, P.W., 2004b. Late Pleistocene/Holocene slip rate of the Zhangye thrust (Qilianshan, China) and implications for the active growth of the northeastern Tibetan Plateau. *Tectonics*, 23 (6), doi: 10.1029/2004TC001653.

Hilley, G.E., Mynatt, I., and Pollard, D.D., 2010. Structural geometry of Raplee Ridge monocline and thrust fault imaged using inverse Boundary Element Modeling and ALSM data. *Journal of Structural Geology*, 32: 45–58.

Howle, J.F., Bawden, G.W., Schweickert, R.A., Finkel, R.C., Hunter, L.E., Rose, R.S., and von Twistern, B., 2012. Airborne LiDAR analysis and geochronology of faulted glacial moraines in the Tahoe-Sierra frontal fault zone reveal substantial seismic hazards in the Lake Tahoe region, California-Nevada, USA. *Geological Society America Bulletin*, 124: 1087–1101.

Hugenholtz, C.H., Wolfe, S.A., and Moorman, B.J., 2007. Sand-water flows on cold-climate eolian dunes: Environmental analogs for the eolian rock record and Martian sand dunes. *Journal of Sedimentary Research*, 77: 1–8.

Hunter, L.E., Howle, J.F., Rose, R.S., and Bawden, G.W., 2009. The "Polaris Fault": A previously unmapped fault discovered using LiDAR near Martis Creek Dam, Truckee, CA. *Seismological Research Letters*, 80: 305.

Hunter, R.E., Richmond, B.M., and Alpha, T.R., 1983. Storm-controlled oblique dunes of the Oregon coast. *Geological Society of America Bulletin*, 94: 1450–1465.

Irvine-Fynn, T.D.L., Barrand, N.E., Porter, P.R., Hodson, A.J., and Murray, T., 2011. Recent High-Arctic glacial sediment redistribution: A process perspective using airborne Lidar. *Geomorphology*, 125: 27–39.

Jessop, D.E., Kelfoun, K., Labazuy, P., Mangeney, A., Roche, O., Tillier, J.L., Trouillet, M., and Thibault, G., 2012. LiDAR derived morphology of the 1993 Lascar pyroclastic flow deposits, and implication for flow dynamics and rheology. *Journal of Volcanology and Geothermal Research*, 245–246: 81–97.

Jimenez, J.A., Maia, L.P., Serra, J., and Morias, J., 1999. Aeolian dune migration along the Ceara coast, north-eastern Brazil. *Sedimentology*, 46: 689–701.

Jones, K.L., Poole, G.C., O'Daniel, S.J., Mertes, L.A.K., and Stanford, J.A., 2008. Surface hydrology of low-relief landscapes: Assessing surface water flow impedance using LIDAR-derived digital elevation models. *Remote Sensing of Environment*, 112: 4148–4158.

Kereszturi, G., Procter, J., Cronin, S.J., Németh, K., Bebbington, M., and Lindsay, J., 2012. LiDAR-based quantification of lava flow susceptibility in the City of Auckland (New Zealand). *Remote Sensing of Environment*, 125: 198–213.

Kieffer, S.W., 1971. Shock metamorphism of the Coconino sandstone at Meteor Crater, Arizona. *Journal of Geophysical Research*, 76: 5449–5473.

Klinger, R.E., and Piety, L.A., 2000. Late quaternary tectonic activity on the Death Valley and Furnace Creek Faults, Death Valley, California. In: U.S. Geological Survey Digital Data Series 058: *Geologic and Geophysical Characterization Studies of Yucca Mountain, Nevada, A Potential High-Level Radioactive-Waste Repository.*

Kondo, H., Toda, S., Okumura, K., Takada, K., and Chiba, T., 2008. A fault scarp in an urban area identified by LiDAR survey: A Case study on the Itoigawa–Shizuoka Tectonic Line, central Japan. *Geomorphology*, 101: 731–739.

Krabill, W.B., Thomas, R.H., Martin, C.F., Swift, R.N., and Frederick, E.B., 1995. Accuracy of airborne laser altimetry of the Greenland ice sheet. *International Journal of Remote Sensing*, 16: 1211–1222.

Lan, H., Martin, C.D., Zhou, C., and Lim, C.H., 2010. Rockfall hazard analysis using LiDAR and spatial modeling. *Geomorphology*, 118: 213–223.

Lancaster, N., 1995. *Geomorphology of Desert Dunes*, Routledge, London, 312 p.

Lato, M., Kemeny, J., Harrap, R.M., and Bevan, G., 2013. Rock bench: Establishing a common repository and standards for assessing rockmass characteristics using LiDAR and photogrammetry. *Computers and Geosciences*, 50: 106–114.

Lato, M.J., and Vöge, M., 2012. Automated mapping of rock discontinuities in 3D lidar and photogrammetry models. *International Journal of Rock Mechanics and Mining Sciences*, 54: 150–158.

Le Gall, B., Authemayou, C., Ehrhold, A., Paquette, J.-L., Bussien, D., Chazot, G., Aouizerat, A., and Pastol, Y., 2014. LiDAR offshore structural mapping and U/Pb zircon/monazite dating of Variscan strain in the Leon metamorphic domain, NW Brittany. *Tectonophysics*, 630: 236–250.

Lin, Z., Kaneda, H., Mukoyama, S., Asada, N., and Chiba, T., 2009. Detection of small tectonic–geomorphic features beneath dense vegetation covers in Japanese mountains from high-resolution LiDAR DEM. *Seismological Research Letters*, 80: 311–312.

Lin, Z., Kaneda, H., Mukoyama, S., Asada, N., and Chiba, T., 2013. Detection of subtle tectonic–geomorphic features in densely forested mountains by very high-resolution airborne LiDAR survey. *Geomorphology*, 182: 104–115.

Liu, W., Dong, P., Liu, J.B., and Guo, H.D., 2012. Evaluation of three-dimensional shape signatures for automated assessment of post-earthquake building damage. *Earthquake Spectra*, 29: 897–910.

Livingstone, I., Wiggs, G.F.S., and Weaver, C.M., 2007. Geomorphology of desert sand dunes: A review of recent progress. *Earth-Science Reviews*, 80: 239–257.

Matsu'ura, T., Kimura, H., Komatsubara, J., Goto, N., Yanagida, M., Ichikawa, K., and Furusawa, A., 2014. Late quaternary uplift rate inferred from marine terraces, Shimokita Peninsula, northeastern Japan: A preliminary investigation of the buried shoreline angle. *Geomorphology*, 209: 1–17.

Mitasova, H., Drake, T.G., Harmon, R.S., and Bernstein, D., 2004. Quantifying rapid changes in coastal topography using modern mapping techniques and GIS. *Environmental and Engineering Geoscience*, 10: 1–11.

Mynatt, I., Hilley, G.E., and Pollard, D.D., 2007. Inferring fault characteristics using fold geometry constrained by Airborne Laser Swath Mapping at Raplee Ridge, Utah. *Geophysical Research Letters*, 34: L16315.

Narteau, C., Zhang, D., Rozier, O., and Claudin, P., 2009. Setting the length and timescales of a cellular automaton dune model from the analysis of superimposed bed forms. *Journal of Geophysical Research*, 114: F03006.

Necsoiu, M., Leprince, S., Hooper, D.M., Dinwiddie, C.L., McGinnis, R.N., and Walter, G.R. 2009. Monitoring migration rates of an active subarctic dune field using optical imagery. *Remote Sensing of Environment*, 113: 2441–2447.

Ni, N., Chen, N., Chen, J., and Liu, M., 2016. Integrating WorldView-2 imagery and terrestrial LiDAR point clouds to extract dyke swarm geometry: Implications for magma emplacement mechanisms. *Journal of Volcanology and Geothermal Research*, 310: 1–11.

Nichols, K.K., Bierman, R.R., Foniri, W.R., Gillespie, A.R., Caffee, M., and Finkel, R., 2006. Dates and rates of arid region geomorphic processes. *GSA Today*, 16: 4–11.

Perron, J.T., Kirchner, J.W., and Dietrich, W.E., 2009. Formation of evenly spaced ridges and valleys. *Nature*, 460: 502–505.

Ping, L., Narteau, C., Dong, Z., Zhang, Z., and Courrech du Pont, S., 2014. Emergence of oblique dunes in a landscape-scale experiment. *Nature Geoscience*, 7: 99–103.

Ramos, N.T., Tsutsumi, H., Perez, J.S., and Bermas, P.P., 2012. Uplifted marine terraces in Davao Oriental Province, Mindanao Island, Philippines and their implications for large prehistoric offshore earthquakes along the Philippine trench. *Journal of Asian Earth Sciences*, 45: 114–125.

Regmi, N.R., McDonald, E.V., and Bacon, S.N., 2014. Mapping quaternary alluvial fans in the southwestern United States based on multiparameter surface roughness of lidar topographic data. *Journal of Geophysical Research: Earth Surface*, 119: 12–27.

Reitz, M.D., Jerolmack, D.J., Ewing, R.C., and Martin, R.L., 2010. Barchan-parabolic dune pattern transition from vegetation stability threshold. *Geophysical Research Letters*, 37: L19402.

Richter, A., Faust, D., and Mass, H.-G., 2011. Dune cliff erosion and beach width change at the northern and southern spits of Sylt detected with multi-temporal LiDAR. *CATENA*, 103: 103–111.

Rubin, D.M., 1990. Lateral migration of linear dunes in the Strzelecki Desert, Australia. *Earth Surface Processes and Landforms*, 15: 1–14.

Rubin, D., and Hesp, P.A., 2009. Multiple origins of linear dunes on Earth and Titan. *Nature Geoscience*, 2: 653–658.

Sabins, F.F., Jr., 1987. *Remote Sensing Principles and Interpretation* (2nd edition), Freeman, New York, 449 p.

Sampson, C.C., Fewtrell, T.J., Duncan, A., Shaad, K., Horritt, M.S., and Bates, P.D., 2012. Use of terrestrial laser scanning data to drive decimetric resolution urban inundation models. *Advances in Water Resources*, 41: 1–17.

Saye, S.E., van der Wal, D., Pye, K., and Blott, S.J., 2005. Beach–dune morphological relationships and erosion/accretion: An investigation at five sites in England and Wales using LIDAR data. *Geomorphology*, 72: 128–155.

Schulz, W.H., 2007. Landslide susceptibility revealed by LIDAR imagery and historical records, Seattle, Washington. *Engineering Geology*, 89: 67–87.

Scott, D.H., and Trask, N.J., 1971. *Geology of the Lunar Crater Volcanic Field, Nye County, Nevada*. Geological Survey Professional Paper 599-I. United States Government Printing Office, Washington, DC.

Shaw, E.W., 1911. High terraces and abandoned valleys in western Pennsylvania. *The Journal of Geology*, 19: 140–156.

She, J., Zhang, Y., Li, X., and Feng, X., 2015. Spatial and temporal characteristics of snow cover in the Tizinafu watershed of the western Kunlun Mountains. *Remote Sensing*, 7: 3426–3445.

Sherrod, B.L., Brocher, T.M., Weaver, C.S., Bucknam, R.C., Blakely, R.J., Kelsey, H.M., Nelson, A.R., and Haugerud, R.A., 2004. *Holocene fault scarps near Tacoma, Washington, USA*. Geology, 32: 9–12.

Sieh, K.E., and Jahns, R.H., 1984. Holocene activity of the San Andreas fault at Wallace Creek, California. *Geological Society of America Bulletin*, 95: 883–896.

Silva, P., Goy, J., Zazo, C., and Bardaji, T., 2003. Fault-generated mountain fronts in southeast Spain: Geomorphologic assessment of tectonic and seismic activity. *Geomorphology*, 50: 203–225.

Sloss, C.R., Westaway, K.E., Huang, Q., and Murray-Wallace, C.V., 2013. An introduction to dating techniques: A guide for geomorphologists. In: (J. Schroder, ed.) *Treatise on Geomorphology*, Elsevier Inc., Amsterdam, The Netherlands, pp. 346–369.

Spinetti, C., Mazzarini, F., Casacchia, R., Colini, L., Neri, M., Behncke, B., Salvatori, R., Buongiorno, M., and Pareschi, M., 2009. Spectral properties of volcanic materials from hyperspectral field and satellite data compared with LiDAR data at Mt. Etna. *International Journal of Applied Earth Observation and Geoinformation*, 11: 142–155.

Staley, D.M., Wasklewicz, T.A., and Blaszczynski, J.S., 2006. Surficial patterns of debris flow deposition on alluvial fans in Death Valley, CA using airborne laser swath mapping data. *Geomorphology* 74: 152–163.

Székely, A., Zámolyi, A., Draganits, E., and Briese, C., 2009. Geomorphic expression of neotectonic activity in a low relief area in an Airborne Laser Scanning DTM: A case study of the Little Hungarian Plain (Pannonian Basin). *Tectonophysics*, 474: 353–366.

Taniguchi, K., Endo, N., and Sekiguchi, H., 2012. The effect of periodic changes in wind direction on the deformation and morphology of isolated sand dunes based on flume experiments and field data from the Western Sahara. *Geomorphology*, 179: 286–299.

Tarquini, S., and Favalli, M., 2011. Mapping and DOWNFLOW simulation of recent lava flow fields at Mount Etna. *Journal of Volcanology and Geothermal Research*, 204: 27–39.

Tarquini, S., Favalli, M., Mazzarini, F., Isola, I., and Fornaciai, A., 2012. Morphometric analysis of lava flow units: Case study over LIDAR-derived topography at Mount Etna, Italy. *Journal of Volcanology and Geothermal Research*, 236: 11–22.

Tong, J., Déry, S., and Jackson, P., 2009. Topographic control of snow distribution in an alpine watershed of western Canada inferred from spatially-filtered MODIS snow products. *Hydrology and Earth System Sciences*, 13: 319–326.

Vianello, A., Cavalli, M., and Tarolli, P., 2009. LiDAR-derived slopes for headwater channel network analysis. *CATENA*, 76: 97–106.

Volker, H.X., Wasklewicz, T.A., and Ellis, M.A., 2007. A topographic fingerprint to distinguish alluvial fan formative processes. *Geomorphology*, 88: 34–45.

Wallace, R.E., 1975. The San Andreas fault in the Carrizo Plain-Temblor Range region, California. In: (J.C. Crowell, ed.) *San Andreas Fault in Southern California: A Guide to San Andreas Fault from Mexico to Carrizo Plain*. California Division of Mines and Geology Special Report, volume 118, pp. 241–250.

Wallace, R.E., 1977. Profiles and ages of young fault scarps, north-central Nevada. *Geological Society of America Bulletin*, 88: 1267–1281.

Wallace, R.E., 1991. The San Andreas Fault System, California. U.S. Geological Survey, Professional Paper 1515. United States Government Printing Office, Washington DC.

Wallace, R.E., and Schulz, S.S., 1983. Aerial views in color of the San Andreas fault, California. U S Geological Survey Open File Report 83-98.

Wasson, R., and Hyde, R., 1983. Factors determining desert dune type. *Nature*, 304: 337–339.

Winstral, A., and Marks, D., 2002. Simulating wind fields and snow redistribution using terrain-based parameters to model snow accumulation and melt over a semi-arid mountain catchment. *Hydrological Processes*, 16: 3585–3603.

Woolard, J.W., and Colby, J.D., 2002. Spatial characterization, resolution, and volumetric change of coastal dunes using airborne LiDAR: Cape Hatteras, North Carolina. *Geomorphology*, 48: 269–287.

Zachariasen, J., 2008. Detail mapping of the northern San Andreas Fault using LiDAR imagery. Final technical report of National Earthquake Hazards Reduction Program, pp. 1–47.

Zhang, D., Narteau, C., and Rozier, O., 2010. Morphodynamics of barchan and transverse dunes using a cellular automaton model. *Journal of Geophysical Research*, 115: F03041.

Zhang, K., Qu, J., and An, Z., 2012a. Characteristics of wind-blown sand and near-surface wind regime in the Tengger Desert, China. *Aeolian Research*, 6: 83–88.

Zhang, D., Narteau, C., Rozier, O., and Courrech du Pont, S., 2012b. Morphology and dynamics of star dunes from numerical modelling. *Nature Geoscience*, 5: 463–467.

Zielke, O., Arrowsmith, J.R., Ludwig, L.G., and Akciz, S.O., 2010. Slip in the 1857 and earthquakes along the Carrizo Plain, San Andreas fault. *Science*, 327: 1119–1122.

Zielke, O., Arrowsmith, J.R., Ludwig, L.G., and Akciz, S.O., 2012. High-resolution topography-derived offsets along the 1857 Fort Tejon earthquake rupture trace, San Andreas fault. *Bulletin of the Seismological Society of America*, 102: 1135–1154.

Index

197